Robert M. Fineman

A License to Heal...
A License to Steal

A License to Heal... A License to Steal

Copyright © 2009 by Robert M. Fineman

All rights reserved. Printed in the United States of America. No part of this book may be used or reproduced in any manner whatsoever without written permission except in the case of brief quotations embodied in critical articles or reviews.

Library of Congress Cataloging-in-Publication Data

Fineman, Robert M.

A License to Heal... A License to Steal: A doctor's observations and conclusions about the medical profession in the United States

ISBN 978-0-578-03488-1

1. Physicians. I. Fineman, Robert M. II. Title

The history of medicine is not the testament of idealistic seekers after health and life; no more than the history of man is more glorious than a catalogue of selfish and brutish unreason shot spasmodically with sanity.

Gordon S. Ostlere, aka, Richard Gordon (b.1921)

Contents

Dedication	ix
Disclaimer	xi
Photographs	xiii
Acknowledgments	xv
Definitions: Medicine, Good, Bad, and Malignant	xvii
Preface	xix
Introduction	1
1 A la famiglia!	5
Eddie (1914 - 1969)	5
Fran (1918 - 2004)	9
Joe (b.1942) and Yale (1951 - 2004)	10
Me (b.1945)	12
2 A Village Called Mount Airy	15
Scouts	15
Public School	17
Other Opportunities in Mount Airy & Philadelphia	18
Temple University	19
Examples Worth Following	22
3 An Exercise in Fantasy	25
Oath of Hippocrates (460 - 370 BCE)	25
Oath of Maimonides (1135 - 1204 CE)	26

4   Achieving Dreams and Climbing Mountains   31

    1966 - 1972: MD-PhD Student, State University of New York, Downstate Medical School, Brooklyn   32

    1972 - 1974: Pediatric Internship and Residency, Duke University Medical Center, Durham, North Carolina   37

    1974 - 1977: Post-doctoral Fellowship in Human Genetics, Yale University Medical School, New Haven, Connecticut   44

5   Medicine and the Medical Profession - my experiences with the good, the bad, and the malignant   48

    Why Did We Become Physicians?   48

    What Kinds of Individuals Became Physicians?   49

    Where Did We Become Physicians?   50

    1977 - 1990: Assistant, Associate, and Full Professor of Pediatrics, University of Utah Medical School, Salt Lake City   50

        The Saga of Dr. Kurveh   73

        The Saga of Dr. Mamzer   87

        The Saga of Dr. Goniff   108

        The Saga of the Deans and the VP for Health Sciences   119

        The President's Saga   129

    1991 - 1994: Director, Maternal-Infant Health and Genetics, Washington State Department of Health; and Clinical Associate Professor, University of Washington School of Public Health and Community Medicine, Seattle   151

1994 - 2000: Medical Consultant, Office of
Maternal and Child Health, Washington State
Department of Health; and 1994 - 2002: Clinical
Professor, University of Washington School of
Public Health and Community Medicine,
Seattle   152

2002 - Present: Clinical Professor, Department
of Pediatrics, University of Washington Medical
School, Seattle   157

6   What Are the Obligations of Physicians, and Why?   162

Glossary   171

Appendix A: Physicians' Prayer, Oaths, and Codes   174

Appendix B: Reasons Why We Became Physicians
at the End of the 20th Century in the United States   191

Appendix C: A *Deseret News* Newspaper Article
and Two University of Utah Committee Reports   193

    Document 1: 9 August 1991, "Two Doctors
    File Bias Suits Against U" *The Deseret News*   193

    Document 2: 4 December 1989, *From the Chair*
    of the UU Retention, Promotion, and Tenure
    (RPT) Standards and Appeals Committee to the
    President of the University: Report of the
    Appeal of Dr. Robert Fineman   194

Document 3: 12 March 1990, *From the Chair of the UU Academic Freedom and Tenure Committee (AFTC) to the President of the University: Report and Recommendations of the AFTC on the Complaint of Dr. Robert Fineman* **199**

Dedication

As we express our gratitude, we must never forget that the highest appreciation is not to utter words, but to live by them.

John F. Kennedy (1917 - 1963)

I dedicate this book to my parents, Edward and Frances Fineman. Hardly a day goes by that I do not think of you and try to live the way you raised Joe, Yale, and me. May your memory be a blessing to everyone who knew you, and may you rest in peace.

Edward and Francis Fineman, 1968

Disclaimer

> *A season is set for everything, a time for every experience under heaven:*
> *A time for being born and a time for dying,*
> *A time for planting and a time for uprooting the planted;*
> *A time for slaying and a time for healing,*
> *A time for tearing down and a time for building up;*
> *A time for weeping and a time for laughing,*
> *A time for wailing and a time for dancing;*
> *A time for throwing stones and a time for gathering stones,*
> *A time for embracing and a time for shunning embraces;*
> *A time for seeking and a time for losing,*
> *A time for keeping and a time for discarding;*
> *A time for ripping and a time for sewing,*
> *A time for silence and a time for speaking;*
> *A time for loving and a time for hating;*
> *A time for war and a time for peace.*
> Ecclesiastes 3:1-8

There have been a lot of books written about medicine and the medical profession but few, if any, are like this one.

Except where stated otherwise, all of the thoughts, comments, opinions, and conclusions expressed in this book are mine. Some of its content will expose certain institutions and unnamed individuals to derision and, presumably, this will cause anger and resentment. Suffice it to say, nothing was written out of malice, cynicism, spitefulness, condescension, an ax to grind, or self-aggrandizement.

In virtually every instance, enough time has passed during which the final outcome of the events and issues described here can be fully validated and appreciated in their context. For all these reasons, I have tried very hard to be honest, discreet, and as reasonable as possible while taking into consideration:

> *The reasonable man adapts himself to the world. The unreasonable one persists in trying to adapt the world to himself. Therefore, all progress depends on the unreasonable man.*
> George Bernard Shaw (1856 - 1950)

Some people may say after reading this book that I should have gone with the flow. To them I say I am very proud I did **not** go with the flow, and instead behaved the way my brothers and I were expected to behave in our parents' home; that is, according to the following philosophical precepts:

> Freedom of choice (Deuteronomy 30:19) - people have the ability to choose their actions **and** we are both accountable and responsible for the choices we make.

> and

> You shall rebuke (Leviticus 19:17) - the obligation to be a critic when you see individuals, a group of individuals, or society making terrible mistakes.

Photographs

Edward and Francis Fineman, 1968	ix
Fishing on the Hoh River, 2005	3
Yale, Joe, and Bob Fineman, 1988	14
Eagle Scout award ceremony, 1962	16
Neil Howard Ettinger, 1980	18
Dr. Hazel Mabel Tomlinson, circa 1940s	20
Volunteering in Israel, 2002	23
Bob and Bonnie Fineman, 1968	24
Participating in a physiology experiment, Downstate Medical School, 1966	33
Skinning hogs in North Carolina, 1974	41
University of Utah, 1982	58
Dr. Garth Myers, 1980	61
Dr. Don Summers, 1982	62

Acknowledgments

Gratitude is not only the greatest of virtues, but the parent of all the others.
                    Cicero (106 BCE - 43 BCE)

My family, teachers, Scout leaders, friends, patients and their families, and others taught me that it takes families and villages/communities to nurture and support children and adults. Therefore, it is an honor and a privilege for me to thank and acknowledge everyone who served as a positive role model and helper to me and/or my family. Since there are so many of you, and because I am fearful I could leave out even one of your names and thereby hurt someone's feelings, I will not attempt to list all of your names here. To all of you I say I cannot imagine how much more stressful and difficult our lives would have been without your love, your help, and your support.

Definitions: Medicine, Good, Bad, and Malignant

Medicine / a noun / according to "Medicine," Microsoft Encarta Online Encyclopedia, 2007:

1: *the science and art of diagnosing, treating, and preventing disease and injury*
2: *the goals of medicine are to help people live longer, happier, more active lives with less suffering and disability*
3: *a business*

With all due respect to the authors of Encarta, # 3 above should be listed as # 1, because for more than 200 years medicine has been more of a business than a profession **[G]*** in the U.S. A 1979 ruling by our Federal Trade Commission, supported by a U.S. Supreme Court decision in 1982, declared the practice of medicine to be a business rather than a profession.

Good / an adjective / for example:

1: the Oath of the Boy Scouts of America
> *On my honor I will do my best to do my duty to God and my country and to obey the Scout Law; To help other people at all times; To keep myself physically strong, mentally awake, and morally straight.*

2: the Law of the Boy Scouts of America
> *A Scout is trustworthy, loyal, helpful, friendly, courteous, kind, obedient, cheerful, thrifty, brave, clean, and reverent.*

(* **NOTE:** Please refer to the Glossary, page 171, wherever a **[G]** appears in this book for additional definitions and/or descriptions of specified words or terms.)

Bad / an adjective / for example:
1: immature
2: simple-minded
3: child-like

Malignant / an adjective / see Webster's Revised Unabridged Dictionary, 1913:
1: *disposed to do harm, inflict suffering, or cause distress*
2: *actuated by extreme malevolence or enmity*
3: *virulently inimical*
4: *bent on evil*
5: *malicious*

Preface

> *The ultimate measure of a man is not where he stands in moments of comfort and convenience, but where he stands at times of challenge and controversy.*
> Martin Luther King, Jr. (1929 - 1968)

Physicians play a very important role in the lives of individuals, families, and communities. Their role involves medical, ethical, psychological, social, legal, financial, and other attributes. Sooner or later each of us, a family member, a friend, or someone we know will interact with a physician.

The interactions of physicians with patients and their families, other physicians and allied healthcare providers, and others vary considerably; that is, from outstanding to awful. There are many reasons for this including *doctorhood*, which I define as the quality or character of a doctor(s). For the most part, understanding the attributes of doctorhood and the medical profession in the U.S., and coping with and/or surviving their deficiencies are what this book is about.

This book is also about my 35+ year career in medicine, university-based academic medicine, and public health. It is not a novel or an in-depth scientific study but, for the most part, a collection of personal reminiscences that took place from the mid-1960s to 2006 at some of America's most prestigious university medical centers. It was written from a bottom-to-top perspective using a combination of narrative, descriptive, didactic, and emotional styles.

In the initial, autobiographical chapters I describe who I am and why I wanted to be a doctor since I was 14-years-old. The remainder of the book describes what was good, bad, and/or malignant about the MDs and PhDs I have known - with a special emphasis on: those who educated/trained doctors, the important lessons I learned about doctorhood and the medical profession, and what will need to happen in the future to improve both professionalism among doctors and healthcare in the U.S.

Obviously, others will have perspectives different than mine because the subject matter of this book is complex, and also because of our different personality types. That is, each of us collects, filters, analyzes, and interprets information differently based, in part, on our knowledge, attitudes, skills, behavior, experiences, and biology/genetics.

My intentions in writing this book are purely beneficent. I hope its publication will help protect and preserve the well-being of individuals, families, and communities who interact with doctors as patients, students, and/or colleagues. In addition, it may help others obtain justice and/or compensation when they have been subjected to the unethical and/or illegal actions of doctors who, in most cases, have taken an oath to do good and not harm others.

Credibility is an extremely important issue when writing a memoir, especially one that is critical of the actions of others. Therefore, I used in several places letters, memos, committee reports, and newspaper and magazine articles, to document my observations and the correctness of my conclusions.

Only a few names are mentioned in the book, including some relatives, teachers, mentors, and friends. This was done, in part, to protect the innocent and the guilty. The good deeds and actions described here, as well as the unethical, immoral, and amoral ones, speak for themselves.

I did not write this book in haste. I have been writing it on-and-off for almost two decades. Virtually all of the interactions and/or situations are ones I personally witnessed. I did not exaggerate, and I tried to the best of my ability to be honest, truthful, accurate, and fair. I have no hidden agendas.

As noted above, it was absolutely essential for me to describe in the initial parts of this book relevant information about my upbringing, education, experiences, and biology/genetics, so it would be very clear where I am coming from when I make statements like: the terms "medical profession" and "corporate integrity" are, for the most part, oxymoronic; the difference between God and a lot of doctors is God doesn't think He's a doctor; and doctors have a license to steal and many have chosen to do so. Many doctors succumb to competing vocational and personal conflicts of interest on a regular basis, thereby giving the rest a bad name because they have acted immaturely, egotistically, selfishly, entitled, greedily, and/or destructively.

Finally, traits like fidelity, piety, humility, integrity, and ethical decision-making among doctors who not only educate medical students and others, but who also administer physician oaths, are not nearly as common as most people think or would like to believe. Doctors and others educate/train students to become doctors, not ethical human beings - not that it would make a difference if they did try to influence students to be or to become ethical.

Robert M. Fineman
Seattle, Washington
August 2009

Introduction

Know Thyself

English translation of words inscribed on the oracle-shrine of Apollo at Delphi, Greece (6th century BCE)

Please permit me to introduce myself, including my:

Basic philosophy:	*If I am not for myself, who will be for me? But if I am for myself alone, what am I? And if not now, when?"* Hillel the Elder (c. 70 BCE - 10 CE)
Biggest challenge:	Staying "cool" when dealing with frustrating situations
Favorite author:	Niccolo Machiavelli (1469 - 1527)
Favorite book:	*The Prince* (c. 1515)
Favorite movies:	*Moonstruck* (1987) and *Field of Dreams* (1989)
Favorite question:	*Am I my brother's keeper?* Genesis 4:9
Favorite retreat:	The Hoh River on the Olympic Peninsula in Washington State, preferably fishing for salmon and steelhead
First job:	Newspaper boy (*Philadelphia Bulletin*) for almost five years
Fondest memory:	Our (Bonnie and my) wedding day, 15 June 1968

Indulgence:	Traveling
Inspiration:	My family, teachers, Scout leaders, friends, and patients, and two people I never met, Albert Einstein (1879 - 1955) and Martin Luther King, Jr. (1929 - 1968)
Major peeves:	Duplicity, greed, and hypocrisy
Passion:	Helping others
Perfect day:	One spent with our family and friends
Proudest moment(s):	The births of our three children and five grandchildren
Virtues:	Passion, perseverance, and resiliency

Throughout my life I sometimes had to be a politician above and beyond the call of duty; for example, when I was confronted with differences of opinion, perceptions, and/or expectations between my bride and my mother - or, as is said in *Yinglish*, a combination of Yiddish **[G]** and English, "*Oy, vey iz mir* (Oh, woe is me); *why did this happen to me, Lord?*"

Other times, I didn't feel I had to be a super politician and a people pleaser. In part, that is another important reason why I wrote this book; that is, to explain how, when, and why I did <u>not</u> do things at work the way my co-workers and colleagues wanted them done.

Finally, if I had it to do all over again, I would live my life pretty much the way I did because the conundrums I faced made for an interesting, exciting, and occasionally nonconformist and confrontational life I am thankful and proud to say appears to be ending happily-ever-after.

Fishing on the Hoh River, 2005

Chapter One

A la famiglia!

> *American families have always shown remarkable resiliency, or flexible adjustment to natural, economic, and social challenges. Their strengths resemble the elasticity of a spider web, a gull's skillful flow in the wind, the regenerating power of perennial grasses, the cooperation of an ant colony, and the persistence of a stream carving canyon rocks. These are not the strengths of fixed monuments but living organisms. This resilience is not measured by wealth, muscle, or efficiency, but by creativity, unity, and hope.*
>
> Ben Silliman (b.1952)

When bad things happen to resilient families, they usually process their feelings, accept reality, and move on with a sense of optimism. Unfortunately, not all families show maximum resiliency under stress. In fact, their solutions (scapegoating, abuse, separation, and/or divorce) sometimes make matters worse.

Resiliency, it should also be noted, is not an either/or phenomenon. Individuals and families can be more or less resilient at any time, and they may vary in their coping capacity from time to time. Such was the case with me and my family when I was a child.

Eddie (1914 - 1969)

> *He that wasteth his father, and chaseth away his mother, is a son that causeth shame, and bringeth reproach.*
> Proverbs 19:26

Relevant to the nature and scope of this book, pearls of wisdom my father told my brothers and me, in a very direct and frank manner, included:

> *Stay in school and learn as much as you can for as long as you can.*
>
> *Participate in Scouts, sports, and as many other activities as possible in the community.*
>
> *Be a person of your word: if you say you are going to do something, do it; or if you make a promise, keep it.*
>
> *There are three things I want you to remember when you grow up: be independent, especially financially; know the difference between right and wrong; and know the difference between cause and effect.*
>
> *Try to do things right the first time so you won't have to look back later and regret what you did in the past.*
>
> *If you are going to be rebellious, you better know exactly who and/or what it is you are rebelling against.*
>
> *Be careful when dealing with others because in this world there are too many mamzers, kurvehs, and goniffs [in Yiddish: a mamzer is a bastard - my dad used the term to describe evil people; a kurveh is a prostitute - he used the term to describe unethical people who sell themselves for material gain; and a goniff is a thief].*
>
> *Never do anything that will bring shame on our family's name.*

Finally, my dad wanted his sons to exhibit real grit and be over-achievers, not under-achievers. He was deathly afraid of wasted potential because he was always talking about opportunities he had missed, and I'm sure he wanted his sons to have grit because grit helps leave as little as possible to chance.

My dad's old-school, college-of-hard-knocks words of advice were on target more than 90% of the time. His advice and the demands he made were clear and simple, and he didn't want to hear any excuses. Interestingly, in today's permissive, I have a right to do pretty much anything I like culture, I presume most people would disagree with my assessment, and many would think my father was overbearing and/or a tyrant. Wrong! What he was trying to do was turn boys into men of character or, as they say in Yiddish/German, *menschen*. The easiest way for a person to accomplish this is through personal responsibility, as opposed to control over one's environment/others.

****

Unfortunately, in regard to his health and well-being, my dad did not always follow his own advice and, in some instances, external circumstances made it impossible for him to do so. Thus, he suffered much hardship and illness during his short life. During the last several years of his life he couldn't walk 25-30 yards without stopping to catch his breath. He died a few hours after having a below-knee amputation of one of his legs; that is, after a kind of surgery he vowed he would never have.

Environmental, psychosocial, and biological (hereditary) factors had a great influence on me, and other members of my immediate and extended family. For example, in 1930 at the age of 16 and at the beginning of the Great Depression, my dad dropped out of high school to work in a factory. His pay was *bupkis* ('beans" in Yiddish, but here it means nothing), but he had no choice because he had to pay for his food, clothing, and shelter. However, dropping out of school caused him to be unfulfilled, angry, frustrated, and bitter later in life.

The sudden unexpected death of my father's father from a heart attack at age 48, in 1937, was also devastating to my dad and his family. There have been numerous people on my father's side of the family who have suffered and/or died prematurely because of metabolic syndrome **[G]**, including my father, his father, my father's younger brother (who died suddenly at age 36), my younger brother, and at least two of my paternal first cousins. Some died quickly while others, like my father and my younger brother, suffered for a long time before they died.

I watched my father, from age 42 - 52, have at least four significant heart attacks. During the last eight years of his life, he was not able to work because he was in and out of hospitals regularly due to poorly controlled insulin-dependent diabetes mellitus, painful diabetic peripheral neuropathy (nerve disease), unstable angina, heart attacks, and chronic congestive heart failure. A man who was very independent and an excellent athlete was reduced to total dependence on others, with an accompanying loss of social status, by age 47.

Our family was on a Social Security assistance program for disabled adults during the last eight years of my dad's life, and that caused him to be depressed, abrasive, angry, and to develop a mercurial temperament. I never had the feeling my father felt optimistic during the last years of his life, and it was obvious that in different ways everyone in our family was significantly affected by his illnesses and disability.

Finally, when my blue-collar father used to say in Yiddish to my brothers and me a major problem with the world is there are too many *mamzers*, *kurvehs*, and *goniffs*, I would tell him that, unlike his less educated blue-collar colleagues, my co-workers in the future were going to be ethical and caring because they would be highly educated doctors. With a chuckle, I have occasionally thought of building a fishing pond in an Iowa cornfield, a Pond of Dreams if you will, because I would really love the "old man" to come back so we could go fishing, and I could apologize to him and tell him how wrong I was. I know now there is no association between higher education, including a higher education that emphasizes ethics, and higher ethical standards or practice.

Fran (1918 - 2004)

My son, hear the instruction of thy father, and forsake not the law of thy mother.
Proverbs 1:8

Relevant pearls of wisdom my mother gave to my brothers and me included:

If you live long enough, you'll see everything.

Leave the world a better place than you found it.

If you can't say something nice about a person, you shouldn't say anything at all.

I might as well give my money away while I am alive because I can't take it with me. [The only money my mom had to give away was the money my brothers and I, and our wives, gave her after dad died.]

My mom was a hard-working and very humane person. She was warm, affectionate, caring, friendly, easy to like, supportive, soft spoken, courteous, and considerate of others. She was genuinely interested in the welfare of others, generous with her time, energy and resources, encouraging, and reassuring. She was popular and people sought her out for friendship. In fact, mom tried her hardest <u>not</u> to disagree with anyone because she was a "people pleaser." I know she had a great memory because until she was in her eighties she could readily remember everything and everyone she agreed with.

Before my dad became ill, mom was the Cub Scout den mother who introduced me to Scouting. Later, she worked at a variety of full-time sales, secretarial, and bookkeeping jobs when my father had to work part-time, after he developed diabetes and had a second heart attack. Looking back, my mom was actually a very resilient person, who I believe could have lived a better life if she spent more time trying to please herself and less time trying to please others.

Joe (b.1942) and Yale (1951 - 2004)

> *A mensch is a decent, responsible person with admirable characteristics; someone to admire and emulate; someone of noble character. The key to being "a real mensch" is nothing less than character, rectitude, dignity, a sense of what is right, responsible, and decorous.*
> Leo Rosten (1908 - 1997)

Joe is 2 ½ years older and Yale was 6 ½ years younger than me. For much of my life I wondered how three brothers could be so different. Joe became a futures/stock broker, and Yale was a musician and a university-based music librarian/administrator. As the years went by, I realized I was wrong and, in fact, all three of us shared the same basic characteristics.

I was a lot closer to Joe when we were growing up, and he had a bigger impact on my life than Yale, most likely because we were closer in age. To help our family when we were youngsters, Joe delivered the *Philadelphia Bulletin* newspaper every day after school, babysat for neighbors' kids, and mowed lawns and shoveled snow on the walkways of neighbors' homes. Almost all the money he made was given to our parents. He also worked his way through college waiting tables; washing dishes, pots and pans; and being a cook at children's summer camps.

Because of Joe, I ended up doing many of the same things. Thus, the two of us were pretty much financially self-supporting by the time we were 14-years-old. Joe and I, and I think our parents, too, were not always happy with the types of jobs we had, but we did them anyhow and with a minimal amount of fuss. They were jobs that had to be done so we could survive as a family - like what our dad did during the Great Depression, except Joe and I didn't have to drop out of school and work full-time, thanks in large part to a Social Security program for which we will always be grateful.

Joe went on to get his bachelor's and master's degrees. He was the first person in our immediate family to graduate from college. When he was in college, some of the money I earned was used to help pay his tuition. Later on, when I was in medical school, he gave back the money. Our family was a team working together to fulfill needs and accomplish dreams.

As for Yale, he was much more rebellious than Joe and me and, unfortunately, he began smoking cigarettes as a teenager. Early on, he did poorly in school and our parents hoped he would be a "late bloomer." In fact, that was what happened. Yale went on to earn a bachelor's and two master's degrees. Unfortunately, he was diagnosed with metabolic syndrome when he was in his late thirties, including insulin-dependent diabetes mellitus and, later on, significant coronary artery disease requiring stents and heart surgery. He died on his 53rd birthday of lung cancer, may he rest in peace. Ironically, he quit smoking cigarettes 12 years before he developed lung cancer.

Shortly after he was diagnosed with lung cancer in March 2004 he said to me, "I have a lot to live for." He was right, and I will never forget those words. I only wish he would have realized them when he was younger and smoking cigarettes.

As mentioned above, I thought for most of my life my brothers and I were very different: a futures/stock broker, a musician/music librarian, and a physician. Then, several years ago I asked Joe what he did when he went to work. His answer, which really surprised me, was "I protect the clients from the firm;" and then he proceeded to tell me how brokerage firms were not always on the up-and-up with their clients, and what measures he took to protect them.

Later on, when Yale died and I learned what his colleagues wrote about him, I was equally proud because they confirmed my thoughts about Yale:

> *Yale will be remembered by his colleagues as a vivid presence, dedicated in equal measure to librarianship, service, scholarship, and musicianship - and as a loyal friend who touched numerous lives across the country.*

In summary, both of my brothers turned out to be real *menschen*; that is, highly ethical, educated, productive, and respected men who my father would have been, and my mother was, very proud.

Me (b.1945)

> *No man is a good doctor who has never been sick himself.*
> A Chinese Proverb

Regardless of man or woman, this Chinese proverb is true. In addition, I believe a person has an increased probability of becoming a good doctor if she/he has grown up with a close relative or friend who was very sick for a long time. The experience is like a hot poker that indelibly burns itself into your soul.

When I was 14-years-old I told my mother for the first time I wanted to be a doctor. We were in the kitchen of our home washing the dinner dishes. Her initial response was, "Where are we going to get the money for you to go to medical school?" I said don't worry, we will find the money and where there's a will (and I didn't mean an inheritance) there is a way. She laughed for a few seconds, then tears welled up in her eyes and she apologized for her response. I told her not to be upset, and that her response was perfectly normal because I didn't know either.

I decided I wanted to be a doctor primarily because of my father's illnesses and my desire to prevent similar suffering in others. The more he suffered, and as a consequence the more our family suffered, the more determined I became. In order to accomplish my goal of becoming a doctor, I worked hard at home, in school, and in the community. Only occasionally did I allow peer pressure, television, dating, the desire to drive a car as a teenager, or anything else get in my way of becoming a good doctor - which is what I hope I am.

I didn't know any doctors up close and personal when I was growing up in Philadelphia, and it never occurred to me to do informational interviews with physicians I didn't know to help me understand doctorhood and the medical profession. Occasionally, I watched television shows like *Ben Casey* (1961 - 1966) and *Doctor Kildare* (1961 - 1966). While I knew they were not real doctors, I thought when I was a teenager that *Casey* and *Kildare* were exhibiting typical physician conduct. In fact, they were inspirational role models for me and the kind of doctor I hoped to be and work with in the future.

Now, as an adult, when I get up in the morning and look in the mirror, I know I am and have always been about 60% Eddie and 40% Frances Fineman. My personality and outlook on life are the result of my parents' genes, half Eddie's and half Fran's, of course, the total environment in which I grew up, and how I responded to both. I think of myself as a survivor of a stressful childhood, and I presume most people would say, "Aren't we all?"

I am a firm believer a person's moral compass and ethical framework are pretty much determined by the time they are 18-years-old, unless there are very extenuating circumstances; for example, a near death experience. A higher education has little or no positive effect on one's ethical standards or moral compass. In fact, the opposite may be true. With a higher education, a person has an opportunity to learn more sophisticated skills that can enable their feelings of entitlement, selfishness, and greed.

I know my personality is not perfect, but I have tried hard to be a *mensch* because that was my parents' greatest wish, and I wanted them to be proud of me.

Yale, Joe, and Bob Fineman, 1988

My brothers and I had parents who loved and deeply cared about us. During our youth, however, I don't think we fully understood or appreciated how much because we grew up in a home where there was not a lot of laughing and joy. This was most likely due to my dad's illnesses and our parents' different personality-types. During the best of times Fran was like salt and Eddie was pepper, but during the worst of times they were like fire (Eddie) and water (Fran). Thus, I think it would be accurate to say that despite the different ways of handling tremendous health and economic difficulties there were always evidences of resiliency in our family because, as was often said in the movie *Moonstruck* (1987), *"A la famiglia!"* was what really counted.

Finally, never in my life did I think I would live to an old age - not when I was a teenager and not when I had a stent placed in my right coronary artery at age 55, the same age my father was when he died. I learned as a child, and re-learned later as a doctor, that life can be short and sometimes very harsh. Thus, despite the traditional Yiddish adage, *Man plans and God laughs,* I have thought, since age 14, everyone should plan ahead and give each day a solid effort because tomorrow, who knows, we may not be here. And, rather than end this chapter on a down note, I want to wish all of us a truly heartfelt *"L' Chaim"* (To Life); may we all live to 120 in good health!

Chapter Two
A Village Called Mount Airy

It takes a village to raise a child.
     An African proverb

Our family moved to the Mount Airy section of Philadelphia in1951, when I was in the 1st grade. I lived there until I graduated from Temple University in 1966. Mount Airy was a community that promoted the growth and development of its children through its many outstanding opportunities. And, as previously mentioned, my parents strongly encouraged me to participate in as many of them as possible.

Scouts

Cub Scouts and Boy Scouts were my childhood substitute for religious training. My family was not religious, at least not outwardly, when I was growing up in Mount Airy. I did not have much formal religious training, nor did I have a *Bar Mitzvah* **[G]**. We celebrated some of the major Jewish holidays at home and, except for a brief period of about 2-3 years, we were not members of a synagogue.

I became a Cub Scout at age 8 and a Boy Scout at age 12. Because of weekly Scout meetings, weekend overnight campouts, Scout summer camp, and other Scout special events, I probably repeated the Scout Oath and Scout Law at least fifty times a year; so much so, they became a reality to me.

The Boy Scout Oath and Law were the Ten Commandments of my childhood, and I believe in them today as much as I did back then. It is very important to note that the Oath and Law, like the Ten Commandments, stress the obligations of a Scout, not his rights. In September 1962, I became an Eagle Scout. Shortly thereafter, I was elected into the Order of the Arrow by my fellow Scouts. My parents were present at both award ceremonies, and I know they were proud of my accomplishments.

Eagle Scout award ceremony, 1962
(I'm the tall guy in the back.)

Many of the skills I learned in Scouts I still use today; for example, hiking, swimming, boating, fishing, cooking, safety, and public speaking. But, even more important were the ethical guidelines, attitudes, and behaviors that Scouting taught me; for example, citizenship in the home, community, and nation. I've tried very hard to live by those guidelines and attitudes all my life and, if I were asked to pick the four areas that influenced my upbringing the most, they would be my family, Scouts, school, and friends.

Public School

The public schools I attended were very good. Class trips to the Gettysburg National Military Park, Franklin Institute and Planetarium, Philadelphia Art Museum, Philadelphia Zoo, and other places, were not uncommon and a lot of fun. In junior high school, I received free French horn lessons including a French horn, and I also participated in the school band and orchestra.

In high school, I lettered in track and football. Chuck Bednarik of the Philadelphia Eagles, the last man to regularly play both offense and defense in the National Football League, was one of my childhood heroes because I played center and, occasionally, linebacker. Football, and I could say many of the same things about band and orchestra, taught me the benefits of teamwork, selflessness, sportsmanship, discipline, enthusiasm, intelligence, focus, a commitment to excellence, and "playing hurt" within reason.

As for my teachers, two of them, Marilyn Appel in biology and Neil Ettinger in chemistry, had a profoundly positive effect on my decision to become a doctor. Marilyn and Neil (it took me years to call them by their first names, at their insistence) were everything I wanted and needed in high school teachers. They were knowledgeable, dedicated, and skilled, and also tough and demanding, but fair. I have stayed in touch with them for more than 45 years. In addition, one of my high school physical education teachers was indirectly responsible for me meeting Bonnie, my bride of 40 years, in the summer of 1964 when she and I worked at his children's summer camp in the Catskill Mountains in New York. How fortunate I was to have so many teachers and classmates who were very helpful and friendly.

Neil Howard Ettinger, 1980

In January 1963, I graduated 4th in a class of about 270 students from Germantown High School. I was awarded a full tuition Temple University scholarship and a partial tuition Philadelphia Public Board of Education scholarship. Unbelievably/thankfully, Temple allowed me to keep both scholarships, which meant I had extra money to pay for books, laboratory fees, and daily living expenses.

Other Opportunities in Mount Airy & Philadelphia

Mount Airy had good, readily accessible public playgrounds and libraries, free art lessons for children at a local public park, and a very large, well-attended concert and fireworks display at Temple University Stadium every July 4th. In addition, there were the unbelievable Big 5 (Temple, University of Pennsylvania, La Salle, St. Joseph's, and Villanova) basketball games at the Palestra (on the University of Pennsylvania campus); and also Philadelphia's finest cuisine: hoagies, cheese steaks, pizza, TastyKakes, soft pretzels with mustard, black cherry wishniak soda, and the Good Humor man and his ice cream truck. The food may sound like prototypical junk food, but going to the Palestra for a Big 5 doubleheader, the whole building rocked and rolled during the games, with mustard on pretzels, is probably as close to Nirvana as I'll ever get. Growing up in Mount Airy was like growing up in a wonderful orchard. I won't go as far as saying it was the Garden of Eden, but all I had to do is reach up to pick the fruit. What more could I ask for!

Temple University

Temple University was a "blue-collar" commuter school of more than 20,000 students when I went there from 1963 - 1966. From its beginning in 1884, Temple sought to educate students exactly like me. It was big, urban, and racially and culturally diverse, much to my liking. The science departments where I spent most of my time had good teachers and, because of the two scholarships I had, it was free.

Dr. Hazel Tomlinson was my freshman chemistry professor. She was challenging and nurturing in equal proportion, patient, caring, and compassionate. Thanks to Neil Ettinger, my inspirational high school chemistry teacher, I got A's in freshman chemistry.

Dr. Hazel Mabel Tomlinson, circa 1940s

In my sophomore year, I began working as a part-time undergraduate assistant for Dr. Tomlinson. I made sure her chemistry labs were well supplied with the proper chemicals and I checked out the equipment, prepared the "unknowns" used in the labs to test students, and graded tests. The pay was good, and I enjoyed working for her until I graduated in June 1966 because I learned more about chemistry from her, and I was able to help other students at the same time.

After I was accepted by the Jefferson Medical College in Philadelphia and the State University of New York, Downstate Medical School in Brooklyn, early in the spring of 1966, I told Dr. Tomlinson the good news. She said she was very happy for me, and then she asked me if she could come to my home to talk to my parents. When I asked her what she wanted to talk about she responded, "Medical school."

To the day I die, I will never forget what happened next. It was during Passover when she came to our home. This little lady in her early-sixties, about 5'5" tall, with gray curly hair and weighing no more than 115 pounds, sat down at our dining room table and ate dry matzo and water. For whatever reason, she refused to eat or drink anything else my mother offered. She started off by saying she had talked to me a month or so previously about medical school costs, and she knew it was a problem for me. Then she said, "I will loan Bob the money for medical school tuition, room and board, interest free, for as long as he needs it." My parents and I were speechless and even now, when I think about Dr. Tomlinson, I get a lump in my throat and tears in my eyes.

Her visit to our home lasted less than 20 minutes and ended with a handshake. My parents and I thanked her profusely, but she said it was not necessary. Many years later I found out why. She had done the same thing for other students, quietly and with no personal recognition. That was the kind of person she was. Fortunately for me, in my sophomore year of medical school I was able to obtain a fellowship that paid for my tuition, room, and board. After Bonnie and I got married in June, 1968, we paid Dr. Tomlinson back the money she had loaned me.

Strangely enough, she and I share the same birthday - May 15th. Sadly, she has passed away and every year on our birthday I say a little prayer for her, quietly and without any fanfare - just the way she would want it.

****

Since I majored in chemistry at Temple, most of the courses I took were in chemistry, biology, physics, and math - and that was a big mistake. I should have taken more courses in the humanities and especially in the history of medicine, abnormal psychology, political science, and medical law, anthropology, ethics, and economics to prepare myself for becoming a future member of the medical community or vocation - as opposed to the medical profession.

Examples Worth Following

After I graduated from college in 1966, one of my mother's brothers who I liked very much, said to me, "Congratulations, Bob, you are a self-made man." I responded, with respect, that I was not a self-made man and I had a family and a community that really cared for, supported, and nurtured me.

In appreciation for what so many people did for us when we were growing up, Bonnie and I have tried hard to emulate the positive role models in our lives. For example, we have loaned money interest free to college students, had several students live in our home, and we created an endowed scholarship fund for needy students attending Temple University, in honor of my parents. Any money I receive from the publication of this book will be added to that scholarship fund.

We helped resettle Soviet Jewish refugees during the 1980s, and in years 1999 and 2002 we performed volunteer work in Israel. When our older daughter, Elyse, was a Peace Corps volunteer in the mid-1990s, Bonnie and I worked with her - actually Elyse did most of the work, all Bonnie and I did was help obtain the money - to build a small junior/senior high school in rural Paraguay.

From 1980 – 1982, I was president *Congregation Kol Ami* in Salt Lake City, and from 2002 - 2004 I was president of *Congregation Shevet Achim* on Mercer Island, Washington. I have been a Scout leader, a volunteer for the March of Dimes Birth Defects Foundation, and a member of Rotary.

Volunteering in Israel, 2002

Bonnie and I did these things not only because we wanted to emulate our mentors, but also because of what my mother often said, "You should leave the world a better place than you found it."

Bonnie and Bob Fineman, 1968

Chapter 3
An Exercise in Fantasy?

> *There is a disconnect between idealized physician conduct and actual medical practice. Recent efforts ... to create a charter for medical professionalism [are] unfortunately an exercise in fantasy.*
> Nicholas Regush (1946 - 2004)

Before I went to medical school, I knew very little about the character of doctors and physician conduct. I did know that the vast majority of medical students in the U.S., upon graduation, took a sacred oath to do good, and not harm their patients and others.

To me, physician oaths were/are like the Boy Scout Oath, except physician oaths have been around a lot longer. And, because most of the Scouts I knew tried hard to live up to the Scout Oath and Law, I thought the same was true for physicians and the oaths we took.

Oath of Hippocrates (460 - 370 BCE)

I swear by Apollo the physician...I will keep this Oath and this covenant....

I will follow that system of regimen which, according to my ability and judgment, I consider for the benefit of my patients, and abstain from whatever is deleterious and mischievous.

I will give no deadly medicine to any one if asked, nor suggest any such counsel; and in like manner I will not give to a woman a pessary to produce abortion. With purity and with holiness I will pass my life and practice my Art….

Into whatever houses I enter, I will go into them for the benefit of the sick, and will abstain from every voluntary act of mischief and corruption; and, further from the seduction of females or males….

Whatever, in connection with my professional practice or not…in the life of men…I will not divulge, as reckoning that all such should be kept secret.

While I continue to keep this Oath unviolated, may it be granted to me to enjoy life and the practice of the art, respected by all men, in all times! But should I trespass and violate this Oath, may the reverse be my lot! [I can't remember if I took this oath upon graduation from medical school in 1972, or the Oath of Maimonides, below.]

Oath of Moses Maimonides (1135 - 1204 CE)

The eternal providence has appointed me to watch over the life and health of Thy creatures. May the love for my art actuate me at all times; may neither avarice nor miserliness, nor thirst for glory or for a great reputation engage my mind; for the enemies of truth and philanthropy could easily deceive me and make me forgetful of my lofty aim of doing good to Thy children.

May I never see in the patient anything but a fellow creature in pain.

Grant me the strength, time and opportunity always to correct what I have acquired, always to extend its domain; for knowledge is immense and *the spirit of man can extend indefinitely to enrich itself daily with new requirements.*

Today he can discover his errors of yesterday and tomorrow he can obtain a new light on what he thinks himself sure of today. Oh, God, Thou has appointed me to watch over the life and death of Thy creatures; here am I ready for my vocation and now I turn unto my calling. [See Appendix A for examples of other physician oaths and codes.]

In 2002, Dr. Edmund Pellegrino wrote in the Medical Journal of Australia (in an article entitled, Medical commencement oaths: shards of a fractured myth, or seeds of hope against a dispiriting future?), "To erase the principles of the medical oath entirely from our consciousness would be to make medicine no more than a commercial, industrial or proletarian enterprise." According to Dr. Pellegrino, there are several reasons why oath-taking has persisted and should continue to persist in the future. That is, an oath: is a solemn promise made on a solemn occasion at which time medical graduates publicly declare their dedication to certain distinguishing moral commitments. It sets the profession apart and it declares that those who take it are committed to something beyond self-interest. Finally, it is a reminder of the continuity of a profession whose roots are in antiquity, has a seed of hope within it, and reminds physicians of their high calling.

While others may disagree with Dr. Pellegrino and his rationale for physician oath-taking, I not only agree with him, but I would add physician oaths are comparable to the Ten Commandments and Jews and others, and the Boy Scout Oath and Law to Boy Scouts. The oaths were not created to proclaim the rights of doctors, but their educational, ethical, and spiritual obligations.

Lastly, physician oaths and codes are <u>not</u> recommendations. They were created specifically to remind doctors of human frailties that can affect their attitude and behavior, regardless of how rich, smart, and/or educated we think we are. Therefore, the oaths and codes are obligations meant to remind doctors not to exhibit certain unethical, immoral, and/or amoral attitudes and actions - thereby improving our performance.

Granted, over the years changes in society, science, and the law have constantly raised new ethical issues and challenges to existing perspectives. However, it stands to reason if a significant percent of doctors were not exhibiting unethical attitudes and behaviors on a regular basis for the past 2,400 years, then it wouldn't have been necessary to create, disseminate, and repeat such oaths.

****

It was both a blessing and a curse for people like me to become physicians in the U.S. at the end of the 20th and the beginning of the 21st century. In addition, I do not think it would have made much difference if we had lived in a different time and/or in another place because the outcome would have been the same. I say this because while most of the events or situations described in this book involved me personally, similar occurrences and outcomes were brought to my attention over the years by many doctors, and there is ample information in the literature and the law relevant to our experiences; for example, the previously mentioned 1979 Federal Trade Commission ruling, the 1982 U.S. Supreme Court decision, and other historic examples:

> *The Company and Fellowship of the Surgeons of London, minding only their lucres* [money or profit] *and nothing the profit or ease of the diseased, have sued, troubled and vexed divers* [various or several] *honest persons, as well men as women, whom God hath endured with that knowledge of the nature, kind and operation of certain herbs, roots and waters, and the using and ministering of them to such as be pained with customable diseases, etc., and yet the said persons have not taken any money for their pains or cunning, but have ministered the same to the poor people only, for neighbourhood and God's sake and charity.*
> Statute 35 Henry VIII, c.1. [This statute, written in the early 1500s, protected alternative healthcare

providers from physicians who were continuously harassing them.]

Whereas by the 9th act of the Assembly held the 21st of October, 1639, consideration being had and taken of the immoderate and excessive rates and prices exacted by practitioners in physic [medicine] *and chyrugery* [surgery] *and the complaints made to the then Assembly of the bad consequences thereof. It so happening through the said intolerable extraction that the hearts of divers masters were hardened rather to suffer their servants to perish for want of fit meanes and applications then by seeking reliefe to fall into the hands of griping and avaricious men.* [This Virginia law of 1639, enacted in 1646, basically says a physician could be arrested and taken to court if accused of excessive charges.]

The bane of modern medicine is a merciless commercialism ...
    Dean Lewis, Johns Hopkins
    Hospital, 1937

and

The most important health-care development of the day is the recent, relatively unheralded rise of a huge new industry that supplies health-care services for profit...This new "medical-industrial complex" may be more efficient than its nonprofit competition, but it creates the problems of overuse and fragmentation of services, overemphasis on technology, and "cream-skimming," and it may also exercise undue influence on national health policy. In this medical market, physicians must act as discerning purchasing agents for their patients and therefore should have no conflicting financial

interests. Closer attention from the public and the profession, and careful study, are necessary to ensure that the "medical-industrial complex" puts the interest of the public before those of its stockholders.
                    Arnold Relman (b.1923)

If I had had more formal education in the humanities, law, business, and psychology/psychiatry when I attended college, medical school, or during my post-graduate training, I would have been much better prepared to understand life in general and doctors in particular - especially their motives, opportunities, and means. If I had had more communication early on with others not associated with the medical-industrial complex, I would have identified at a much earlier stage in my career the significant disconnect that has adversely affected the conduct of many physicians for hundreds, if not thousands, of years.

Chapter 4
Achieving Dreams and Climbing Mountains

> *Our lives are inspired by the dreams we have from the earliest stages of our youth. When you combine passion and hard work, then success is always possible. While no road is ever straight, dedication and persistence will always lead you to your dreams.*
>
> Arturo "Arte" Moreno (b.1946)

After graduation from college, my dream consisted of completing my medical and graduate school education and training, becoming a faculty member at a university medical center, getting married, having/raising children, continuing and/or establishing new long-term friendships, and finding ways to help other people, especially children.

Children, especially those with special healthcare, education, and social services needs, have always held a special place in my heart. That's because they are among the most needy and defenseless individuals in our society.

Becoming a seasoned mountain climber, according to discussions I have had with a mountain-climbing friend of mine, and becoming a physician are similar in many ways; for example, staying focused, being persistent, setting lofty goals and achieving them, learning a lot from books, practicing in initially safe settings with a lot of oversight, and applying learned skills in progressively more challenging environments. Also, both are mentally challenging in the sense of being able to employ sound judgment/decision-making in potentially life threatening situations, keeping a "cool" head, anticipating change, doing continuing education, and sometimes pushing yourself beyond your physical and psychological limits.

Lastly, being able to trust your guides, instructors, and fellow climbers is an extremely important attribute when scaling a twenty thousand-plus foot, snow covered peak. There are some people who you can trust with your life, and others you can't. The ones you can trust are the ones you climb with. The ones you can't trust are the ones you don't climb with unless you have a death wish. It is as simple as that. Unfortunately, it's not quite that simple when you are a doctor because you can't always choose the people you work with and you don't always work with people who you can trust - and that's a real shame.

Climbing is, above all, a matter of integrity.
Gaston Rebuffat (1921 - 1985)

1966 - 1972: MD-PhD Student, State University of New York, Downstate Medical School, Brooklyn

I chose to go to medical school in Brooklyn rather than Philadelphia for several reasons: the State University of New York (SUNY) Downstate Medical School had a good MD-PhD program that I knew about when I was applying for admission; tuition for out-of-state residents was about a $1,000.00 a year; Bonnie, my beloved, was a third-year undergraduate, occupational therapy student at Columbia University and living in Manhattan; and since I grew up in Philadelphia I thought it would be both fun and educational to live somewhere else. On the other hand, I had heard rumors that students at Downstate sabotaged other students' laboratory projects and hid reference books in the medical school library to deny others access to the books. I never saw or heard of these things happening while I was there. While there was competition among the students, it was not destructive. In fact, there was a great deal of camaraderie among the students who, like me, lived in the dormitories.

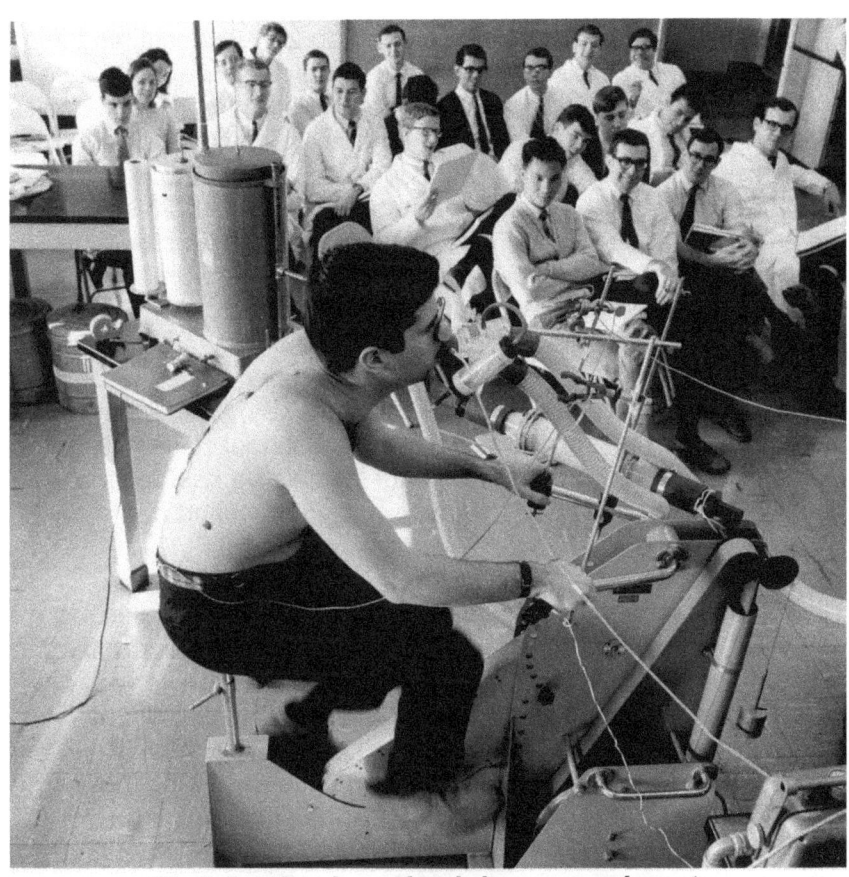

**Participating in a physiology experiment,
Downstate Medical School, 1966**

The first year of medical school was, for the most part, a boring experience because the students were treated like sponges. We were asked to memorize an enormous amount of information, including a whole new vocabulary, and every few weeks a test would come along and the sponges would be wrung out. We never had the opportunity to learn the basic sciences (gross and cellular anatomy, embryology, biochemistry, and physiology) in depth during the first year. What we learned was about a mile wide and six inches deep - not what I would call very exciting. In order to do something different and to make life more interesting and less boring, I ran for a freshman student body officer position.

The freshman class had slightly more than 200 students, and in a non-scientific way I talked individually and briefly to approximately 100 of them in order to find out what they wanted from a class officer. During our conversations, I made it a point to ask the primary reason they wanted to be doctors. About 50 of the 100 said they wanted to be a doctor to make money, and 25 said for the financial security. A definite minority said to help others, to avoid going to Viet Nam, or because their parents wanted them to be doctors. I was very impressed with the honesty of my classmates' answers. When I tried to summarize what I learned from our conversations, I decided not very much since I did not question my classmates in a valid scientific way, and I knew from talking to my professors that each medical school class has its own personality. Therefore, I thought it would be best to store the results of those conversations in the back of my mind and move forward.

Toward the end of the first year of medical school, I applied for and was accepted into the MD-PhD program in the Department of Anatomy and Cell Biology, in part to escape the monotony of med school, and also because I was very interested in pursuing a career in academic medicine, including performing medically-oriented research, teaching, and seeing patients. I also thought the PhD training program would counter-balance my medical school education, and it would open doors for future job opportunities.

The faculty members in the Department of Anatomy and Cell Biology were a terrific group, mostly PhDs and a few MDs. They were good teachers, knowledgeable, always available, and very helpful. In addition, the PhD training program was indeed a right brain, counter-balance to the MD program - the PhD program was about six inches wide and a mile deep. I learned how and how not to perform research, and I specialized in a very narrow field - functional sex reversal in fish. Except for the following experience, and that experience was a harbinger of things to come, I very much enjoyed the PhD side of the MD-PhD combined degree program.

My PhD mentor was a highly published professor who had a reputation for making sure his grad students were well-trained and they almost always got their degrees. In 1970, my mentor asked me if I wanted to be the third or fourth author of a book chapter he was writing. In order for me to be one of the co-authors, he wanted to include in the chapter about a third of my PhD thesis data. Being naïve and not knowing the ramifications of his offer/request, I told him it sounded good to me. I figured it would be an honor to be a co-author of a book chapter.

A few days later, during a conversation with a professor in the Department, I happened to mention the chapter my mentor and I, and two or three others, were co-authoring. He said without hesitation it was not a good idea to allow my thesis data to be included in the chapter because I would not be able to publish it later in a peer-review journal with me being the first author - as it should have been. I went back to my mentor the next day and asked him to leave my thesis data out of the chapter because I intended to publish it in a peer-review journal in the future - which we did. Even though I didn't know if my mentor had made similar requests of other grad students, after that experience I had a gut feeling my thesis mentor was someone I might not be able to trust in the future.

From September 1967 to June 1970, I attended year-round medical and graduate school classes, performed laboratory research, taught human (gross) anatomy, and presented research papers at local and national meetings. In 1969, I presented a paper at the annual American Association of Anatomists' Meeting in Boston. At the meeting a professor of anatomy from a mid-western university, whom I had never met and who had a special interest in human genetics attended the presentation.

After my presentation, he asked me a few questions about my research, and later that day we went to dinner at the Union Oyster House. We talked about a lot of things including my future career plans. He strongly advised me to go into the field of human genetics, adding that over the next 30-40 years genetics would be one of the most exciting areas in all of science and medicine. He was absolutely right, of course, and that meeting/conversation was a major turning point in my career - a meeting that was *beshert*, which in Yiddish means "fated" or "destined."

From September 1970 to June 1971, I experienced regular third-year, medical school inpatient and outpatient clerkships or clinical rotations at several hospitals in Brooklyn. Downstate was a "hands-on" school for medical students during the clinical years; for example, students were able to deliver 15-20 babies or more during our obstetrics and gynecology rotation; we did lots of spinal taps and other procedures; and on Saturday nights we could go to the Kings County Hospital emergency room to participate in the treatment of cardiac arrest patients, patients with severe asthma attacks, accident cases, or to help with the victims of the Battle of Brooklyn - stabbings, gunshot wounds, beatings, and other forms of mayhem.

During the summer of 1971, Bonnie and I lived in Baltimore so I could take a fourth year medical student, elective externship **[G]** at the Joseph Earle Moore Clinic at the Johns Hopkins Hospital. The Moore Clinic was directed by the world famous doctor and *mensch*, Victor McKusick, and it had, arguably, the greatest collection of faculty members and post-doctoral fellows in clinical genetics ever assembled in one place. It was a place where I could kill three birds with one stone: work with two biostatisticians to finish the data analysis and writing of my PhD thesis, evaluate patients with genetic disorders and birth defects under the tutelage of some of the best clinical geneticists in the world, and experience life in another part of the U.S.

Bonnie and I should have bought stock in U-Haul because almost every place we moved over the next several years, U-Haul, my brother, Yale, or Bonnie's brother, Perry, helped us get there!

From September 1971 to May 1972, Bonnie and I lived in Boston where I took additional senior medical student externships/electives at the Boston Children's Hospital and Medical Center. And, once again, I had the opportunity to learn pediatrics and genetics from several top-notch doctors and researchers. In addition, with the completion of the writing and the defense of my PhD thesis in the fall of 1971, I had far fewer school-related deadlines and more time to spend with Bonnie and our first child, Elyse. So, we spent time acting like tourists and took occasional educational day-trips to the historical sites in and around Boston and New England. Life was good. In fact, it was great.

Except for one brief encounter that didn't involve doctors, I don't remember much about graduating from Downstate in 1972. It was a conversation I had with my mother-in-law, bless her heart. She and I were talking, and she asked me what kind of doctor I was going to be. I told her I was going to be a medical geneticist, and one day in the future I would be doing research, teaching, and seeing patients in a medical school - what was referred to back then as a "triple threat." She asked, "You mean you are not going to be a real doctor like a surgeon, or an ophthalmologist, or a radiologist?" I said no. Her response was and still is a classic in our family, and one we still laugh about, "You mean you went to school all those years to be a school teacher?" I thought to myself, "With an MD and a PhD, I thought I was a real doctor."

1972 - 1974: Pediatric Internship and Residency, Duke University Medical Center, Durham, North Carolina

On the road again, just can't wait to get on the road again.
The life I love is makin' music [in my case, medicine] *with my friends, and I can't wait to get on the road again.*
On the road again, goin' places that I've never been, seein' things that I may never see again, and I can't wait to get on the road again.
Willie Nelson (b.1933)

There were several pediatric internship and residency training programs I really liked including the ones at Children's Hospital of Philadelphia and at Duke University. After interviewing at several hospitals, I was very happy when the National Intern and Resident Matching Program matched me with the Duke Pediatrics Program. Duke had a good reputation, I liked the program and the people I interviewed with, and neither Bonnie nor I had lived in the South. Going to back to Philadelphia didn't sound very adventuresome, and after living in Philadelphia, Brooklyn, Baltimore, and Boston, we both realized there was more to life than medicine and my career.

As an aside, when Bonnie and I lived in North Carolina we - she more than me - suffered more culture shock than any time since we met in 1964. In fact, Bonnie could not wait to leave North Carolina, and she literally packed up all our belongings six months before we left, primarily because of the racial and ethno-cultural situations we experienced there. For example, there was an interracial married couple, living in an apartment next door to us, who were verbally abused in supermarkets, restaurants, and other places on a regular basis. One person, a total stranger, referred to their daughter as a "mistake." Another time, when I told one of our "southern neighbors" all the Baptists I knew when I was growing up in Philadelphia were African Americans, he responded, "They're not Baptists, they're niggers." Oops! That was the last time I talked to him.

Actually, the culture shock started before we got to Duke when we were driving south in North Carolina on Interstate-85. We got off the highway north of Durham to see what the countryside looked like. At the end of the exit ramp there was a billboard with a "white knight" riding a horse and the caption, "Welcome to the Heart of Klan Country Welcome to Granville County, North Carolina." Bonnie looked at the sign and said, "There must be a Kentucky Fried Chicken place around here somewhere." I had no idea what she was thinking, and I responded, "I don't think so. That's a Ku Klux Klan sign welcoming us to North Carolina." Bonnie stared at me and said nothing - I could tell she was not amused.

A little further down the road we stopped at a small gas station. The gas station attendant filled the tank of our car, no problem. Yale pulled the U-Haul truck up to the pump. The attendant took one look at Yale and his shoulder-length hair and said, "I don't sell gas to hippy sons-of-bitches." Yale and I looked at each other and we both smiled because we were thinking the same thing. I paid the attendant for the gas he put in our car and then we drove the car and the truck about a quarter-mile down the road. Yale and I switched places, I made a U-turn, drove the truck back to the gas station, got a full tank of gas, and we continued on to Durham. Talk about adventures, welcome to North Carolina!

In fact, the bizarreness of that day was not over. We drove into Durham late in the afternoon and we thought it would be a great idea to get some ice cream. So we stopped at an ice cream bar -

Durham had milk (ice cream) bars back then but no bars that served liquor. Bonnie, in traditional New York fashion, ordered a black and white ice cream soda - chocolate syrup, seltzer, and vanilla ice cream. To make a long story short, the waiter brought Bonnie two sodas, one all chocolate and the other all vanilla. What happened next is something we still joke about to this day. My sweet, loving, kind-hearted bride who was, unbeknownst to me, still unhappy about the white knight sign and the gas station scene said to the waiter with a smile, "I ordered a black and white ice cream soda. What's the matter, are you afraid to integrate them?" I've always thought Bonnie was trying to be funny, but the store manager, who happened to be standing nearby, overheard her comment and he promptly escorted Yale, Bonnie, Elyse, and me to the door, and told us never to come back!

Incidentally, I never saw any racism, ethno-cultural intolerance, or cultural incompetence at the Duke Hospital. On the other hand, there were very few African American medical students, interns, residents, post-doctoral fellows, or faculty members there. However, the same could be said about most of the places where I trained or worked. It was not a good situation then and, to the best of my knowledge, it still isn't today at many university hospitals in the U.S.

****

The first few months of my internship at Duke were difficult because we were in hospital 70-80 hours a week or more; I felt like I was a stereophonic sound system with all its basic parts, however, with a lot of its wires missing; and I was thinking more like a PhD than an MD. This caused me to spend too much time and go into too much detail with each patient I saw. I had a lot of basic medical knowledge in my head; however, I was having trouble connecting it effectively and efficiently to patient care. Fortunately for me, several pediatric faculty members and residents were very helpful and with some extra oversight and suggestions things got a lot better after a couple of months. I will always be indebted to those individuals for their help and understanding during my difficult transition period.

****

There was one pediatric faculty member, Dr. Jim Renuart, who was particularly helpful to me and many other interns and residents. Jim was a pediatric neurologist who was also a gentleman hog and cattle farmer. Whenever I think of the word "mentor" I think of Jim. He was an outstanding clinician and teacher who really cared about his patients, their families, and his students. His pearls of wisdom included: a doctor who takes care of children and who does not have dirt on his or her knees isn't worth a damn - after all, how can you talk to a little kid eye to eye if you are not on your knees; having your ass shot out of the ivory tower on rounds by an intern, resident, or a post-doctoral fellow who spent several hours in the medical school library the night before is one of the greatest rewards a faculty member can attain; never pick up a piglet when the mother pig is around; and stick with me and I will show you how to kosher a hog. The last two pearls were given to me not in the Duke Hospital, but when I went to visit his farm.

One day I arrived at the farm before my mentor/friend, and I happened to pick up and pet a free-roaming piglet in the presence of its mother. What the hell did I know! I was a city boy who knew very little about farms or farm animals. The piglet let out a squeal and its mother came right at me. With the piglet held under my arm I started running across an open field towards a large pile of old, cut-up tree stumps and trunks that was in the middle of the field. At the same time, Jim pulled up in his car, got out, saw me and the pigs, and started yelling, "Drop the pig, drop the pig!" In full stride I dropped the piglet as gently as I could and then climbed to the top of the pile. That mother pig could have bitten off my leg because back then the only thing I knew about pigs was they were not kosher.

Another time, a small group of interns and residents went to the farm to help kill, gut, and skin about a half dozen "top hogs" (200 pounders); and then we took them to Central Carolina Farmers (CCF) to be butchered. Several days later, we went back to CCF to get the meat and something called fatback. Of course, I had never heard of fatback and, besides, Bonnie would not let any of the stuff in our apartment. So, I gave it away to the nurses who worked in the outpatient pediatrics clinic at the hospital.

Skinning hogs in North Carolina, 1974
(Jim Renuart is standing front, left. I'm the guy in the University of Arizona shirt.)

****

During my internship year, spending a minimum of 70-80 hours a week for three months, in hospital, on the inpatient ward service was common place. And, during the three months we worked in the Newborn Intensive Care Unit (NICU) we were expected to be in hospital more than 100 hours a week. It was really insane, however, no one complained and I still haven't figured out why. I presume it was because none of us knew or cared enough to know the difference between persistence and passion versus obsession and neurosis. Of course, having fewer house staff than was needed did save the Department money, and/or maybe everyone including the faculty thought it was a character-building exercise.

I remember one morning when we were making rounds on the ward service a house officer/resident was so sick with gastroenteritis she walked down the hall with intravenous (IV) fluid line in her arm attached to an IV bag/pole on wheels. The interns begged her to go home until she felt better and was no longer contagious, but she would not leave.

Another time, when I was working in the NICU and we were making morning rounds, I fell asleep while sitting on a small metal stool. I fell sideways almost off the stool and hit my head on a plate glass window about 12 inches away. Fortunately, the window was wire-reinforced - otherwise I might have fallen through it. The NICU attending, Dr. George Brumley, who was an excellent physician and an outstanding person, said jokingly, "Bob, I hope the patients got more sleep than you did last night." At the time I thought his remark was funny, but another sleep deprived situation that took place in the NICU several months later convinced me it wasn't.

I was in the NICU for about 25 hours straight, and I was supposed to give a sick newborn a small amount of IV sodium bicarbonate solution. Instead, I drew up in the syringe potassium chloride. Fortunately, the baby was connected to a cardiac monitor. After giving the baby a very small amount of the wrong medication, his heart slowed down by about 10-20 beats per minute. I knew immediately something was wrong. I stopped what I was doing, checked the bottle which was only a few feet away, and realized I was giving the baby the wrong stuff. After what seemed like an eternity, but was probably less than 20 seconds, the baby's heart rate returned to normal and he later went home and did well. That was my first and only patient, to the best of my knowledge, who I gave the wrong medication, caused me to have "brown pants syndrome," and he was a kid I'll never forget.

The pediatric interns and residents at Duke cared a great deal about their patients and their families. However, there was also an unspoken rule among the house staff: <u>No one dies on my shift.</u> For us, it was like a contest to keep the patients alive until the next shift came to work. In fact, it was not good for the patients or their families. Some of the kids were sick beyond belief, and we really felt badly for them. Unfortunately, no one would let one of them die on their shift because the prevailing concept was with an IV, some medications, a feeding tube and/or a ventilator we could keep anyone alive. The hospice concept emphasizing palliative rather than curative treatment, which began in the mid-1970s, was mostly missing.

****

Rather than end on a down note these stories about my internship and residency at "Mr. Duke's Hospital," there are two additional vignettes I need to mention.

It was a quiet December weekend afternoon on the pediatric ward. About 4-5 interns and residents were taking a break, watching an NFL playoff game in our house officers' quarters. The Chairman of the Department of Surgery came on the ward. He was making rounds with his interns, residents, and medical students. It was the usual scene of the head duck followed by all the little ducklings in their white coats, pants, and shoes. Suddenly, deliberately, and without warning he was standing in the door of our quarters yelling, "When I build the new Duke Hospital there won't be any TV sets in the residents' quarters. Get out of here!"

It was like a comedic fire drill with the pediatric house officers running to get out of the room, through a small door, all at the same time. After the Chairman and his ducklings left the ward, we slowly went back to the room to watch the end of the game. Someone asked, "Who was that guy?" Another person said it was the "crazy" Chairman of the Department of Surgery, and we all laughed till it hurt. I think that was the only time I ever saw him, and to this day I still don't know if there are TV sets in the house officer's quarters in the new Duke Hospital.

Last, but not least, I discovered grits and developed a taste for them when I was at Duke. Unfortunately, they tasted somewhat bland to me. So I learned to put Jell-O on them to make them taste better. The cashier of the hospital cafeteria and I became friends, and he used to say to me, "Dr. Fineman, you are the only person in the world who puts Jell-O on top of grits." It's been more than 30 years since I left Duke, and after several Internet searches over the years I think the cashier was right. I still can't find any recipes that mix Jell-O with grits, and that's too bad because when you eat them together it tastes really good.

1974 - 1977: Post-doctoral Fellowship in Human Genetics, Yale University Medical School, New Haven, Connecticut

Following completion of my pediatric internship and residency at Duke, I began a three-year, post-doctoral fellowship in human genetics at the Yale University School of Medicine.

Yale had and still has one of the premier human genetics departments in the U.S. It excelled in just about every facet of genetics research, education, training, and patient-related/clinical expertise. Like the academic institutions where I had been before, it was an honor to be there. I received excellent post-doctoral training at Yale, and I will always be greatly indebted to the faculty members and others there for putting the final touches on my formal education.

During the time I was at Yale, the field of human/medical genetics was just beginning to head down the path of "merciless commercialism" - like what had happened in many other areas in medicine, previously. One of the many things that really impressed me about Yale's Department of Human Genetics was its commitment to academics rather than merciless commercialism. For example, their medium-size cytogenetics (chromosome studies) laboratory was performing more than a 1,000 tests each year. Because of the increased demand for such studies, the laboratory could have increased in size, perhaps by becoming a regional or national laboratory, and performed many more tests. However, this didn't happen because the faculty/leadership in the Department determined the number of tests being performed was sufficient for the genetics program to fulfill its academic mission.

There was one incident involving me when I was at Yale that I still feel badly about. A highly educated father brought his infant son to our outpatient clinic for genetic counseling purposes. The child had Down syndrome, a well known birth defect and intellectual disability syndrome (IQ 40-60).

The first question the father asked me was, "Will my son go to college?" I answered I doubted it very much, and then I went on to ask him a variety of questions about the family history, mother's pregnancy, labor, and delivery, etc. About 20 minutes into the visit, after I had collected a fair amount of medical, social, and previous test results information, the father repeated his question, "Will my son go to college?" I said I doubted it very much, and then I spent the next 30 minutes providing the father with anticipatory guidance information. I made sure I told him to be especially careful not to expect too much from his son because both of them probably would become extremely frustrated. Parents' expectations can play a significant role in their accepting and coping with disabilities in a child with a condition like Down syndrome.

Finally, he asked me the same question again, "Will my son go to college?" At that point I became frustrated and I lost my cool, and responded, "Your son may go to Harvard but he won't get into Yale." I immediately apologized to him for my cruel, uncalled for remark, and I told him I was very sorry for what I said. He didn't respond.

About a week later I was summoned to the office of the Physician-in-Chief of the Yale Hospital. He had received a letter of complaint from the father who was very angry because I told him his son would not be going to Yale. I apologized to the Physician-in-Chief, and I assured him in the future I would not say anything like that again.

This last vignette took place while I was at Yale, however, it did not involve anyone from Yale except me. During my third and final year of post-doctoral training, in mid-November 1976 when I was seeking my first faculty position/job, I was contacted by a doctor from the Baylor University Medical School in Houston. He asked me if I would consider a job opportunity there beginning 1 July 1977. Because Bonnie was in her final month of pregnancy with our third child, I asked him if it would be acceptable if I could delay my interviews at Baylor for a few weeks, until after Bonnie delivered. He said fine, no problem.

About 10 days after Bonnie delivered on November 29th, I went to Baylor to interview and to present a one hour seminar on my research. The interviews were very strange in that we talked about the weather, sports, and everything but the job opportunity there. At the end of day 1 of my visit, after my seminar, I went across the street to the University of Texas Medical Center for a pre-arranged meeting with a medical geneticist friend whom I had met and kept in contact with since my time at Johns Hopkins. During our discussion, he asked me what job at Baylor I was interviewing for. I told him the director of the cytogenetics laboratory. He looked at me with a puzzled look on his face and said, "I don't think so; Baylor hired [Dr. XY] two weeks ago for that position."

The next morning I met with my host physician, the same guy who asked me to come to Baylor to interview for the job opportunity. I asked him about Dr. XY and what job he accepted at Baylor. With an absolutely straight face the host physician admitted there was only one job opening, and it was offered to and accepted by Dr. XY, two weeks previously. I then asked him, "What am I doing here, a week after my wife delivered our third child, and how come you didn't call me and tell me there was no longer a job opportunity for me at Baylor?" His response was, "We wanted to hear about your research, and we thought Baylor would be a good place for you to present it, and to get some feedback about it." I politely told him there were dozens of people at Yale who could critique my research, and I left on an early flight back to New Haven.

When I got back to Yale, I told my Chairman about my misadventure at Baylor. He said, "Bob, the folks at Baylor are good researchers, they just don't know how to recruit new faculty members." He could have said what happened to me at Baylor was an example of blatant dishonesty, or he could have said:

> *Never ascribe to malice that which can be adequately explained by incompetence.*
> Napoleon Bonaparte (1769 - 1821)

> and/or

> *The difference between genius and stupidity is that genius has its limits.*
> Albert Einstein (1879 - 1955)

When I was growing up in Philadelphia, I thought people were basically good. Now I think they are morally and ethically neutral, until proven otherwise. Ultimately, the only part of my dream about becoming/being a physician that did not come to fruition was to work with a large majority of doctors who were competent, and who acted in a highly ethical and humane manner with their patients, families, students, and with each other. This was because I was naïve, or my expectations about doctors, especially those in academe were too high, for both reasons, or:

> *I realized early on that the academy ... [is] dominated by fools, knaves, charlatans and bureaucrats. And that being the case, any human being, male or female, of whatever status, who has a voice of her or his own, is not going to be liked [or accepted].*
> Harold Bloom (b.1930)

Chapter 5
Medicine and the Medical Profession - my experiences with the good, the bad, and the malignant

All the world is a stage, and all the men and women merely players.

William Shakespeare (1564 - 1616)

If Shakespeare had surveyed the doctors I worked with, I think he would have said with a chuckle, "What I see is a soap opera or a three-ring circus, and all the men and women merely players." With this in mind, let me begin this chapter by describing briefly background information relevant to my dream/soap opera/three-ring circus.

Why Did We Become Physicians?

There were many reasons why we became physicians in the U.S. at the end of the 20th century. I explained my reasons earlier. Medicine was a passion for me. I worked 50-55 hours a week during most of my career, and I can honestly say it was an honor and a privilege to be a physician because, in addition to helping people, almost every day was different, challenging, and exciting.

Others had a variety of reasons, noted in Appendix B, for becoming a physician. Some of their reasons were noble and altruistic, some were selfish and self-serving, and some had reasons which were somewhere in between.

What Kinds of Individuals Became Physicians?

Character-wise, in my opinion, every kind of individual became a doctor - from the most ethical, altruistic/patriotic, humane, compassionate, and caring; to the immature, hypocritical, selfish, egotistical, dishonest, self-centered, and greedy; to individuals who were truly psychiatrically impaired. And by psychiatrically impaired I mean people who appeared to have a broad range of over-the-top problems with cognition (perception and interpretation of self, others, and events); affectivity (the range, intensity, lability, and appropriateness of emotional response); interpersonal relationships; and impulse control. Among doctors who seemed to me to be psychiatrically impaired I saw paranoid, schizoid, antisocial, borderline, and narcissistic personality-types; the severely and chronically depressed; and individuals with symptoms of bipolar disease.

The sense of entitlement I've observed in many physicians has never ceased to amaze me. When I was about 10-years-old I began a lifelong fascination with World War II. Over the years, I've read a lot about the Second World War, and I have talked to many WW II combat veterans. If ever a group of individuals deserved to feel entitled, it would be these unsung heroes. In fact, I've never heard one of them express that sentiment; that is, "society owes me." Each said the ones who never came back and the ones who were badly wounded were the real heroes and the ones society owes. If getting torpedoed, bombed, and/or shot at do not create a sense of entitlement, why should sitting in a classroom taking notes or in a library reading medical text books or journals?

Finally, almost all the doctors I have known had/have well above average intelligence quotients (IQ's). Unfortunately, many had ethical, emotional, social, and spiritual quotients that were below average - not only according to my standards, but also according to society's standards. In fact, it was difficult to have a normal conversation over a cup of coffee with many of the university-based physicians I worked with, and there were some who you wouldn't want to shake hands with for fear they would steal the rings off your fingers.

Where Did We Become Physicians?

In the U.S., for about the past hundred years, medical schools and/or academic medical centers have been the primary places where medical and graduate students in the medical sciences, interns and residents, and post-doctoral fellows received their education and training.

1977 - 1990: Assistant, Associate, and Full Professor of Pediatrics, University of Utah Medical School, Salt Lake City

> *Behold! Seven years are coming - a great abundance throughout all the land of Egypt. Then seven years of famine will arise after them and all the abundance in the land of Egypt will be forgotten; the famine will ravage the land. And the abundance will be unknown in the land in the face of the subsequent famine - for it will be terribly severe.*
> Genesis 41:29-31

In July 1977, at the age of 32, I began my first job as a real doctor. I was a junior faculty member, an assistant professor, in the Department of Pediatrics at the University of Utah (UU) Medical School.

I interviewed twice for the UU position during the spring of 1977. During the interview process, I noticed a large plaque in the main entrance of the UU Hospital that eloquently proclaimed the goals of the faculty, staff, and students:

> **RESPECT FOR THE INDIVIDUAL AND SOCIETY**
> *Respect for the dignity and rights of all.*
> **PATIENT**
> *Dedication to provide the highest possible quality of care, delivered with compassion, to preserve and restore health and alleviate suffering*
> **TEACHING**
> *Commitment to an unparalleled educational experience for learning in the health sciences and to provide society with competent, committed professionals*
> **RESEARCH**
> *Recognition that research - the generation, sifting, sorting, and testing of ideas to advance knowledge and better the condition of life - is critical to world progress and a primary purpose of and justification for all centers of higher learning*

Though not mentioned in the plaque, ADMINISTRATION is another major component of medical schools and hospitals. Administrators, such as Division Chiefs, Chairpersons, Deans, Vice Presidents, the President, and others are supposed to be role models and visionaries, who facilitate, integrate, coordinate, and advise faculty, staff, and students. They have to be committed to a maximum degree of flexibility and communication when planning the short and long-term goals and objectives of their programs and organizations. They also have to strive to consistently maintain the following qualities: fairness, integrity, compassion, confidentiality, effectiveness, good judgment, and ethical leadership and decision-making.

The resources needed to achieve the goals of an academic medical center are people, money, time, and space. Far and away the most important resource is people. Excellent administrators, faculty, staff, and students attract other outstanding people. They also attract money that can be used to buy time, and build and maintain outstanding systems, programs, and facilities.

Peer review is another important process of academic medicine. Students, staff, faculty, and administrators are expected to challenge statements and decisions regardless of whether they are dealing with patient care, teaching, research, or administration. Peer review performed openly, honestly, and with respect is absolutely essential for maintaining excellence in every aspect of the academic enterprise.

At that time my romantic, idealistic, and naïve beliefs about academic medicine were further substantiated when I read a copy of the UU *Code of Faculty Responsibility*. This document codified the rights and obligations of the University and its faculty members, and established procedures to assure their observance. In essence, the *Code* stated, in part:

> *The University is not just a corporate body created by law. It is also a community of people associated in activities related to thought, truth, and understanding....*
>
> *The relationship between the University and its faculty should be one of shared confidence, mutual loyalty, and trust. Dealings should be conducted with courtesy, decency, and concern for the personal dignity and shared human values which exist among venturers in the academic enterprise....*
>
> *A faculty member has the legal rights and privileges of a citizen. He or she may not be subject to punishment or reprisal for the exercise of such rights and privileges....*
>
> *In any disciplinary matter, a faculty member has the right to adequate notice, to be heard, and to decision and review by impartial persons and bodies. In disciplinary proceedings involving the possibility of substantial sanctions, a faculty member has a right to due process and peer judgment....*
>
> *Faculty members are entitled to support and assistance from the University in maintaining a climate suitable for scholarship, research, and effective teaching and learning. The University*

> *shall strive to assist faculty members in improving their skills and developing their talents as teachers and scholars....*
>
> *When called upon to serve in administrative posts or on committees* [for example, the University Retention, Promotion, and Tenure (RPT) Standards and Appeals Committee, and the Academic Freedom and Tenure Committee or AFTC], *a faculty member should strive to achieve the legitimate purposes of the University with due consideration for the interests of other persons involved....*
>
> *When a faculty member is engaged in joint research or other professional efforts with colleagues, he or she must exercise reasonable care to discharge his or her agreed obligations to them....*
>
> *Faculty members share the general legal duties of citizenship. A faculty member who violates state or federal law may expect no immunity or special protection by reason of his or her faculty status.*

Every aspect of academic medicine seemed to flow naturally from the goals, philosophical beliefs, and guidelines noted in the plaque and in the *Code of Faculty Responsibility.* It all seemed so simple. A university medical center must have the following four attributes to function optimally: the right people with the right knowledge, attitude, behavior, and skills in the right positions (human resources); a well-defined, rational system; a process that enforces compliance and accountability; and educational processes that ensure a basic understanding of the goals, objectives, values, and processes of the system, and the roles and responsibilities of the individuals who participate in it. During most of the 1980s, the UU Medical School had all but three of these attributes in place. It had a well-defined, rational system on paper, but that was all! What I didn't know was the words in the plaque and the words in the *Code of Faculty Responsibility* were meaningless at the UU Medical School. I didn't know it at the time because I wasn't street smart and because:

> The function of wisdom is to discriminate between good and evil.
> Cicero (106 BCE - 43 BCE)

****

During and immediately after my job interviews at UU and the Utah Department of Health (UDOH), several letters of recommendation about me were sent to the Chairman of the Department of Pediatrics including:

> I am happy to write to support the recommendation that Dr. Robert Fineman be appointed as Assistant Professor at the University of Utah School of Medicine. Dr. Fineman came to our program here at Yale in July 1974 with a strong scientific and clinical background....
>
> During the past two and a half years he has been a trainee in our medical genetics program and has done a fine job. Dr. Fineman has extended his strong medical background to become a perceptive and knowledgeable medical geneticist. He is thorough in his approach to patients, inquisitive to the nuances of disease, and knowledgeable about a broad range of clinical genetic disorders. He has, as you know, a particular interest in patients with chromosome abnormalities and multiple congenital anomaly syndromes, but his knowledge of biochemistry and metabolism is increasingly broad.
>
> In summary, I consider Dr. Fineman to be a good candidate for a junior faculty position. I have no doubt about his ability to launch an independent research effort. He will take his place in a clinical genetics unit effectively and comfortably.

> *Furthermore, and importantly, he has demonstrated excellence in his teaching activities. In this regard I should mention that Dr. Fineman has participated as a discussion leader in our basic genetics course for first year medical students and has received high marks from this very critical audience.*
>
> Dr. Leon Rosenberg, Professor and Chairman, Department of Human Genetics, Yale University

and

> *It was a pleasure for me to have the opportunity to talk to Dr. Robert Fineman for rather long periods of time on both of his visits here. I am extremely impressed with him. He seems to be a very sensitive and humane individual and has been able to sift through the many requests and interviews scheduled for him during those times. He seems to be able to organize his thoughts well and is doing an excellent job mapping out a strategy for his first months here....*
>
> *I look forward to working with Dr. Fineman and expect his presence to improve the health of the citizens of our state.*
>
> Dr. Peter van Dyck, Director, Family Health Services Division, UDOH

In fact, when I was hired, a contract between the UU and UDOH was implemented in which each would pay 50% of my salary and fringe benefits - at least that was what was supposed to happen over the short and long term.

****

I very much enjoyed my job at the UU Medical School during the first several years, and it was obvious my Chairman thought I was doing well because he wrote the following letter of support for my file at my initial, three-year retention action in 1980:

I feel you can take real pride in your accomplishments. You have succeeded in developing an outstanding chromosome laboratory. A genetic counseling team has been developed and a referral out-patient genetics clinic has been established. Your enthusiasm and energy has led to the unusual situation of funding for a new genetics program through the state legislature under conditions when many programs were being reduced and few new programs have been funded. These efforts have made possible the recruitment of a research faculty member who serves as the technical director for the chromosome laboratory and a second medical geneticist [who I have referred to as **Dr. Kurveh** in this book] *who joined our faculty last July....*

All of these efforts point toward the continued development of an outstanding program in Medical Genetics in our Department. As a teacher, house staff has been particularly appreciative of your efforts both as an attending physician and as a consultant. Your presentations have evidenced an excellent fund of information and ability to organize that information in an understandable fashion. The development of the medical genetics course as part of the second year medical school curriculum represented a major effort. I feel that development of the genetics course is a real addition to our curriculum and that you should be commended for your efforts. Any new course represents a major commitment and I am certain that this will be increasingly recognized as the course becomes established.

The development of your research program appears to be evolving in an appropriate fashion. You have completed several studies since joining our faculty and have been effective in working collaboratively with a number of faculty members in investigations which have resulted in several presentations and publications. The potential for increasing productivity in collaboration with other faculty members appears to be excellent and the likelihood of funding for your research programs portends an exciting future. I would also like to commend you for your efforts in continuing to work with individuals under difficult circumstances and in situations where sensitivities are high and people have strong feelings of what is right.

In summary, as your chairman, I strongly support the recommendation of our Department's Promotion, Retention and Tenure Committee that you be retained as an Assistant Professor. I would further like to express my personal appreciation for your efforts in developing our program in medical genetics. You have done an exceptionally fine job in establishing a program and I look forward to increasing excellence in your area of genetics and to the many contributions you will make to the development of our Department.

There was no organized, comprehensive clinical genetics program at the UU Medical School when I started working there in July, 1977. A small chromosome laboratory in the Department of Pediatrics was performing about a 100 studies a year. The laboratory was directed by someone who had relatively little training in medical genetics or cytogenetics, and it had only one technician. I was hired to replace that laboratory director, and this did cause some difficult circumstances. I offered him the opportunity to be the assistant laboratory director, but he refused my offer and went to work in another program in the Department of Pediatrics.

Many other components of combined medical school-wide and statewide clinical genetics program were missing. This was apparent to all concerned. Knowing this, I negotiated, up front, a deal with my future Chairman during my initial job interviews. I would spend 4 - 5 years building a comprehensive medical genetics program at both the medical school and state levels, and then I would be allowed to pursue my research interests at 50% of my work time. The Chairman accepted my proposal. He said, **"No problem. After you build these two programs I promise to protect you so that you can pursue your research interests at 50% of your time, providing you find the money to support your research efforts."**

Within three years, utilizing funding sources from predominantly outside of the University, a newly renovated chromosome laboratory was performing more than 500 studies a year, and it had a new PhD technical director and four technicians. I was its medical director.

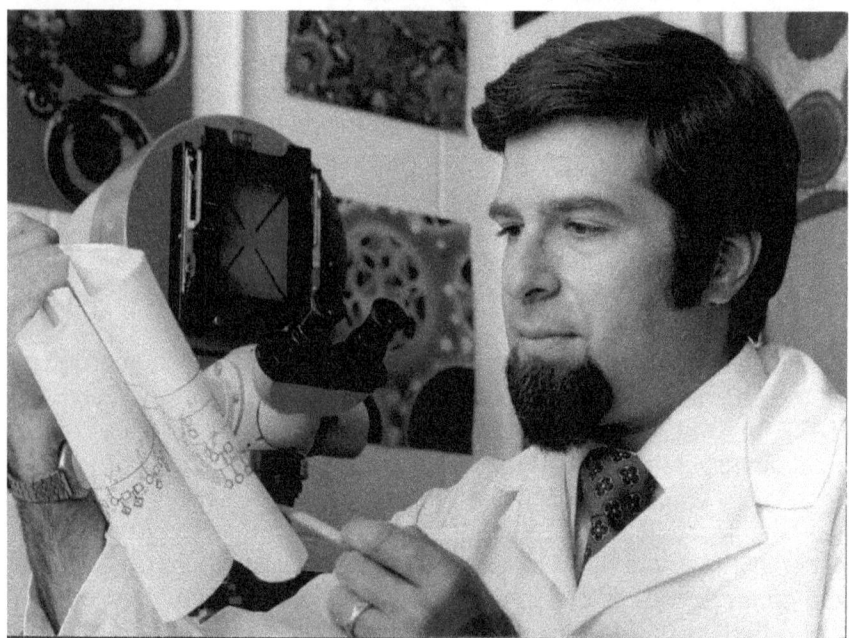

University of Utah, 1982

The chromosome laboratory was also paying taxes that could be used as unrestricted funds by the Chairman of the Department of Pediatrics and the Dean of the Medical School.

At the same time, I worked with Dr. Peter van Dyck and many others to establish a statewide early identification and genetics referral program. The program included inpatient and outpatient genetic/birth defect consultation services at the UU Medical Center and the Primary Children's Hospital/Medical Center, both in Salt Lake City, and in satellite outpatient clinics in Ogden, Vernal, Price, Moab, Blanding, Richfield, Cedar City, and St. George.

In addition, in collaboration with physicians from the UU Departments of Obstetrics and Gynecology, and Radiology, a prenatal diagnosis program was established. A part-time physician in the Department of Pediatrics agreed to oversee a biochemical genetics laboratory on campus.

Lastly, I agreed to oversee an outpatient biochemical genetics/inborn errors of metabolism clinic that treated patients with disorders such as phenylketonuria or PKU **[G]**, and galactosemia **[G]**. This clinic was held at a UDOH facility located next to University Hospital.

These new activities required our rapidly growing Division, with me as the Division Chief, to hire another clinical geneticist, Dr. Kurveh; plus the PhD cytogeneticist and laboratory technicians I mentioned previously, several master's level genetic counselors, and two additional secretaries. Almost no financial support for the Division's rapid expansion came from the Medical School. It all came to the Medical School from the Utah legislature via the UDOH - mostly because of my efforts and the efforts of UDOH personnel - and from monies generated by the chromosome laboratory, seeing patients, and from several successful grant applications I wrote to the Utah Chapter of the March of Dimes Birth Defects Foundation, the Thrasher Research Foundation of Salt Lake City, and to our federal government. I was the Principal Investigator or PI **[G]** of these grants and contracts.

During my job interviews at UU, the Department Chair promised he/the Department would pay for a secretary for the Division of Medical Genetics for two years. Six months after I came on-board he reneged on his promise and he told me I would have to find the money to pay for all of her salary because there were other areas in the Department that needed the money more than Medical

Genetics. When I reminded him that I had submitted several grant proposals he co-signed, which said the Department of Pediatrics would provide in-kind secretarial support, he said he changed his mind and the Division of Medical Genetics would have to come up with the money.

Later on, I obtained copies of financial records which showed that my Chair, over time, modified unilaterally my 50-50 salary and fringe benefits arrangement he had negotiated with the UDOH. Unbeknownst to me or UDOH staff, UDOH ended up paying about 70% and the Department of Pediatrics only about 30% of my salary and fringe benefits. I was very disappointed to learn what my Chair was doing because it was very dishonest, but I learned to cope with it.

****

During my time at UU, I had the opportunity to work with some really outstanding doctors. Like Dr. Jim Renuart at Duke, they were great teachers and they inspired others by their conduct and love of humanity. The welfare of others was their foremost priority, and they were not only well versed in the art and science of medicine, but with life in general.

Dr. Garth Myers was a pediatric neurologist who oversaw the Birth Defects Clinic at the Primary Children's Medical Center from 1962 to 1988. Garth was a down to earth person with no hidden agendas. Everyone knew where they stood with Garth. The first time I met him was during one of my two job interviews, and the first thing he said to me after, "Hello, my name is Garth Myers," was "You're not taking over my position as head of the Birth Defects Clinic." I had no idea why he said this and to be honest I really didn't want to know. I responded, "I have no intention of taking over your Clinic and no one has suggested I do that. All I want to do is work with you." And for more than a decade that's what happened. We worked together not only at the Birth Defects Clinic but also in UDOH-sponsored, children with special healthcare needs, outpatient/satellite clinics all over Utah. Garth cared about the health and well-being of patients and their families more than virtually any other doctor I have ever known.

That's not to say Garth didn't care about his co-workers and others. He did very much, however, the patients and families came first and he ran a top-notch, multi-disciplinary birth defects clinic that included specialists in pediatrics, and pediatric neurology, neurosurgery, orthopedics, ophthalmology, genetics, speech and hearing, nursing, nutrition, and social work. I don't think even one doctor ever refused to participate in the Birth Defects Clinic after Garth asked them to do so. Garth was a real *mensch* and, equally important, he didn't care much for bureaucrats.

Dr. Garth Myers, 1980

Dr. Peter van Dyck was the Director of the Family Health Services Division, UDOH, from the early 1970s to 1992. He was and still is an outstanding physician and administrator and, probably, he did more for Utah's children and families than any other doctor before or since. He had an enormous fund of knowledge, and I learned a great deal from him about the principles and practice of public health including the public health core functions: assessment, policy development, and program implementation - before these functions were described in detail in a 1988 *Institute of Medicine* report entitled, *The Future of Public Health.* Peter was honest.

He had no hidden agendas, and his door was always open. If it were not for Peter many of the family-oriented maternal and child health programs in Utah, including the statewide clinical genetics program, would not have been created.

Dr. Don Summers was the Chairman of the Department of Cellular, Viral and Molecular Biology at the UU Medical School. Don was tall, built like my father, and he had a personality that was similar to my dad's. You knew what he was thinking and where he was coming from because he was an honest, caring, straight-shooter. His laugh was infectious, and he was a great story-teller.

During one of my mother's visits to Salt Lake City I introduced Don to her as my surrogate father. Even though he had multiple sclerosis, Don was very energetic and passionate about almost everything. We talked about science, art, music, literature, travel, history, family life, and even fishing. Don and I were "squash partners." We played twice a week for almost a decade. He loved life and intensely disliked bureaucrats.

Dr. Don Summers, 1982

Regarding bureaucrats, I had a conversation with a Dean of the UU Medical School that I will describe in detail, later. The conversation was about ethical and financial conflicts among physicians working in private practice in Salt Lake City. A few days after my conversation with the Dean I had a conversation with Don about medical school deans. I asked him why anyone would want to be a Dean and, in general, what were they like? I didn't mention my recent conversation with our Dean.

Don asked me, "Bob, don't you remember the story I told you about medical school deans a couple years ago?" Unfortunately, I had forgotten what Don had told me:

> *Arguably, the best Dean the UU Medical School ever had was Dr. John Dixon, a surgeon from Ogden, Utah. One day, Dean Dixon went to an annual meeting of all of the medical school deans in the U.S., and after he came back he had a brief conversation with Don. Don asked the Dean the following question, "John, among the more than 100 med school Deans in the U.S., how many of them are any good?" The Dean thought for several seconds and said, "Six." The next day Dean Dixon and Don ran into each other in the hallway and the Dean said, "Don, I'm sorry I made a mistake yesterday; I should have said five."*

Unfortunately, Dr. Dixon retired from his deanship position in 1976, the year before I arrived at UU. During my 13 ½ years there - I left in December 1990 - there were four Deans. During the decade of the 1990s there were four Deans. Unlike medical school, department chairs at UU who had a position for life, the deanship there, like in many other places, was a revolving door with Deans going in and out all the time. Why? Primarily it was because the Deans at the UU Medical School had little or no power, and they could be ignored or, even worse, devoured by Chairs at any given time. One of the things that made Dean Dixon so special was the way he could make all the lions and tigers, that is, the UU department chairs, sit up on their hind legs like in the circus - and he did it without using a whip and a chair!

As an aside, when Don Summers celebrated his 65th birthday in 1999, almost a decade after I left UU, I sent him and his bride, Ellie, the following E-mail:

Forever Don

Dear Don and Ellie,

The higher I climbed the academic ladder, the less I was able to identify with my colleagues. In fact, at each higher level from high school, to college, to medical school, and so forth, I noted a higher and higher percentage of people who were more interested in themselves than they were in others....

Fortunately, there are always exceptions to such observations and the two of you were among the exceptions. Sorry, Don, I couldn't leave Ellie [who was also a professor at the UU Medical School] out of this E-mail even though it is your birthday. On her 65th birthday, I will send the same E-mail to her with the names reversed, and call it Forever Ellie.

Sure, both of you are smart and successful, but the one thing I will never forget about you is that you cared about people! And it didn't make a difference if it was a secretary, the janitor, a student, or a fellow faculty member....

Your door was always open, you genuinely tried to help people, and you even treated a naïve jerk like me with respect and dignity.

I thank you both from the bottom of my heart. It is really a shame that there are not more people like you in this world. Don, may you live to be a 120 years old, and may we all be there in good health to celebrate it.

Unfortunately, it was not meant to be. Don died a few years later, may he, too, rest in peace. And to this day, occasionally I still ask myself, "What would my parents and/or Don Summers do in a situation like this?"

I also had the distinct pleasure of interacting with other good people in Utah outside the UU and UDOH, including the leadership of the LDS (Mormon) Church, the Utah legislature, Governor's office, Attorney General's office, Utah's federal representatives, March of Dimes, the local newspapers, and many others.

One time the President of the University and I visited the Salt Lake City office of **Senator Orrin Hatch (R-Utah)** to talk about federal legislation regarding newborn screening, prenatal diagnosis, and other aspects of genetics in public health. We met with Senator Hatch for about a half hour and he was wonderful - smart, engaging, and well-spoken with a good sense of humor.

At the end of the conversation, I noticed the Senator was wearing cowboy boots and, of course, I was wearing mine because in addition to being a doctor I wanted to be a cowboy, too. I asked the Senator if he knew the difference between a "real" cowboy and a "drugstore" cowboy. He said no. "It's simple," I said, "The real cowboy has the bullshit on the outside of his boots." Senator Hatch was still laughing after the President and I said our goodbyes and headed for the door.

****

Being a faculty member at UU was not all work and no play. The UU indoor recreation/sports facilities for students, faculty, and staff; its football, basketball, and women's gymnastic teams; and going skiing, fishing, river-running, and sightseeing in Utah's fabulous state and national parks made for many wonderful times. Our Mormon and non-Mormon neighbors were among the best Bonnie and I ever had. The Jewish community was equally outstanding.

In 1981, I presented a paper at a meeting in Jerusalem. On the way back Bonnie and I stopped off to visit her relatives in London, where I bought a couple of expensive Cuban cigars. When we got back to Salt Lake City, I lit up one of the cigars at work and I asked one of our secretaries if she could smell the difference between the twenty cent cigars I occasionally smoked and this very expensive Cuban cigar. She responded, "Dr. Fineman, can you smell the difference between pedigree dog crap and mongrel dog crap? All your cigars smell like dog crap." No doubt about it, a good secretary is worth her or his weight in gold!

Then, there was the time when about two dozen members of the Salt Lake City Jewish community held a "Las Vegas Night" that included illegal, at least in Utah, gambling like card and dice games, and a roulette wheel. At the end of the evening when it was time to cash in our chips, a suggestion was made to donate all of the gambling money to the Primary Children's Hospital and Medical Center. Since I knew the physician-in-chief there, a very kind and friendly Mormon physician, I volunteered to deliver the money to him.

The next day I went to his office and said, "Sir, I have good news for you and I have bad news." He asked, "What's the good news?" I responded, "I have a lot of money in this envelop that two dozen members of Salt Lake's Jewish community want to donate to Children's Hospital." He asked, "So, what's the bad news?" I said, "The money is sin money according to Mormon beliefs. We held a Las Vegas Night last night, with real gambling, and this is the money we used." In a heart-beat he said, "Give me the money!" I asked, "Don't you want to think about it?" "Nope, not even for a second," he responded. We laughed and I gave him the envelope.

****

> *Nothing is more noble, nothing is more venerable than fidelity. Faithfulness and truth are the most sacred excellences and endowments of the human mind.*
> Cicero (106 BCE - 43 BCE)

In the late 1970s and early 1980s, I directed or helped co-direct the *First, Second, Third, and Fifth Annual Birth Defects, Mental*

Retardation, and Medical Genetics Meetings in Salt Lake City. The meetings were co-sponsored by our Division of Medical Genetics, the UDOH, and from a grant I obtained from the Herbert I. and Elsa B. Michael Foundation of Salt Lake City. They took place at first in a UDOH facility and later at different hotels in Salt Lake City, and they were attended by doctors, nurses, social workers, laboratory personnel, and healthcare administrators. The speakers were always nationally known and usually well known by someone in our local speaker selection committee.

One of the meetings was devoted to medical ethics and, in this instance, several of the speakers, while nationally renowned, were not known personally by any of the members of the selection committee or consulting clergy. Who would have ever suspected we would have more difficulties dealing with individuals trained in ethics than any other type of professional we asked to speak at our meetings?

A few weeks after the medical ethics meeting, and after the presenters had received their remuneration checks, I received a letter from one of the presenters who said he was underpaid by 50%. I responded in writing and included a copy of his speaker's contract. I asked him to re-read the contract he signed before the meeting. It showed that the stipend he received was appropriate for a speaker who attended only one day of the two-day meeting. A few weeks later, I was summoned to the office of the President of the University because he had received a letter of complaint from the ethicist/speaker.

When I met with the President, I showed him the signed contract which clearly stated two levels of remuneration, one for speakers who stayed for only one day and one for those who stayed for both days of the meeting. I explained the idea was to provide a financial incentive to speakers who would attend both days of the meeting, which this man didn't. The President agreed with me, and we both laughed in regard to the audacity displayed by this medical ethicist. The President wrote the speaker a letter explaining our position, and we never heard from him again.

Another ethicist who made a presentation at the meeting asked for twice as much money as the other speakers received. Our meeting committee agreed to his request. As with everyone who made a presentation at the meeting, we asked this speaker to give us a hard copy of his presentation because we wanted to publish the proceedings of the meeting in a medical journal. Even though he agreed to give us a copy of his presentation at the time of the meeting, he didn't. When we again asked him for it, he requested one of our secretaries transcribe the tape of his presentation so that he could review and comment on it, and then send the hard-copy back to me.

One of our secretaries spent hours transcribing his presentation but, of course, he never followed through with his promise. In fact, we never should have sent him his remuneration check until we received a hard-copy of his presentation. However, we naïvely thought if you can't trust a medical ethicist to keep a promise, who can you trust?

Finding people in academe who were both knowledgeable and ethical became a huge issue for me as time went by.

In February 1982, my Chairman died from a heart attack at age 49. He and I got along well. He was supportive, for the most part, and he tried to advance my career and the careers of many others whenever he could.

It took about 18 months to hire my Chairman's permanent replacement, a man referred to in this book as **Chairman Mamzer or Dr. Mamzer**. This book probably would not have been written if my first Chairman hadn't died, or if Chairman Mamzer wasn't who he was.

A few months after Dr. Mamzer became Chairman of the Department of Pediatrics it was apparent our relationship was not good. Knowing things were not going well, I sought advice from my friend, Don Summers. I told Don about several interactions I had had with Chairman Mamzer, and then I asked him to explain to me what was wrong. Don smiled and said, "Bob, your Chairman suffers from hubris." Don was right, and at that point I had a strong premonition Chairman Mamzer and I were heading for some very intense and stressful confrontations, primarily because I knew who I was and I had a pretty good idea who he was. After my conversation with Don, I thought about what my father had told me when I was a teenager: *If you are going to be rebellious, you better know exactly who and/or what it is you are rebelling against.*

****

> *If you know the enemy and know yourself, you need not fear the result of a hundred battles. If you know yourself but not the enemy, for every victory gained you will also suffer a defeat. If you know neither the enemy nor yourself, you will succumb in every battle.*
> Sun Tzu (c. 544 BCE - c. 496 BCE)

One of the things Chairman Mamzer and I talked about early in his time at UU was my previous Chairman's promise to allow me to spend 50% of my time performing research after the multi-departmental, medical school-wide and statewide medical genetics programs were created, and after I found the funding needed to support my future research efforts.

Happily, the university and statewide clinical genetics programs were doing very well. While there was room for improvement, the important and many of the less important parts of the programs were in place.

After Chairman Mamzer and I met a few times, I began to realize there was something very different about his eyes. They reminded me of Jack Nicholson's eyes when he leered through the broken doorframe and shouted "Heeeere's Johnny!" in the movie, *The Shining* (1980).

About three months after his arrival, Chairman Mamzer asked for the resignations in writing of all of his division chiefs. I guess he was having similar thoughts about me and my presence because my resignation was one of those he accepted. I presume it was because he wanted to surround himself with weak-personality Division Chiefs who he could easily pressure into doing whatever he wanted.

In any case, my almost seven years of abundance at UU was coming to an end because an era of immorality, selfishness, omnipotence, incompetence, greed, and mindless commercialism was descending on the Department of Pediatrics. Unbeknownst to me, this was occurring with the approval and support of the Dean of the Medical School, the Vice President (VP) for Health Sciences, and the President of the University, who happened to be a physician and a previous VP for Health Sciences at UU.

****

> *If a man can accept a situation in a place of power with the thought that it's only temporary, he comes out all right. But when he thinks that he is the cause of the power - that can be his ruination.*
> Harry S. Truman (1884 - 1972)

In my opinion, about 15-20% of the doctors I dealt with over the years were highly moral/ethical people; about 20-25% were amoral, immoral and/or unethical, and the rest were somewhere in between, depending on which way the wind was blowing that day. In order to document and confirm my observations and conclusions, it was necessary to include in this book actual letters, memos, committee reports, and newspaper and magazine articles. I quoted their words here so there wouldn't be any issues/difficulties understanding what I said versus what they said.

30 December 1983: A letter I received from Chairman Mamzer

> *During our recent meeting, I expressed my concern about your Division's focus and processes. We discussed the division of work in the Division and the coverage of the genetic outreach clinics. In addition, I outlined my concerns over the manner in which your Division is viewed within the institution and the community,* **at least based on my limited inquiries and observations** *[emphasis mine]. With the uncertainty that exists surrounding your "sabbatical" year for next July [1984 to June*

> 1985], *I have decided to replace you as director of the Division of Medical Genetics. [Dr. Kurveh] will take over for you as Division Chief.*

My being a division chief for almost seven years was long enough. I also knew if I continued being a Division Chief I'd have to work closely with Chairman Mamzer, and I didn't want to do that, in part because of his shoot-from-the-hip style as evidenced in his letter above.

What I really wanted to do beginning in July 1984 was write grant proposals and do research during what turned out to be a "part-time" sabbatical year; that is, I had to agree to continue to be the medical director of the chromosome laboratory during the sabbatical, a task that accounted for 33-40% of my work time, because no one else at UU had the ability to oversee the laboratory.

In regard to my other, regular responsibilities (being an attending physician on a pediatric inpatient/hospital service, a medical genetics inpatient consultant at the Children's and University Hospitals, teaching medical genetics to freshman and sophomore medical students, and seeing patients with birth defects and genetic conditions at the Children's Hospital Birth Defects Clinic and the UDOH statewide, children with special healthcare needs outreach clinics), Chairman Mamzer brought into the Division a long-time medical geneticist friend of his, **Dr. CD**, to perform these activities in my place.

14 May 1984: A letter I received from the President of the University

> *Based upon the formal review carried out during this academic year of your accomplishments as a faculty member, and in accord with the applicable provisions of the University and Faculty Regulations, I am pleased to award you tenure [G] as an Associate Professor effective July, 1, 1984.*
>
> *This action reflects the confidence of the University community in your ability and in your*

commitment to the advancement of knowledge in your discipline. You have my best wishes for continuing progress in your career at the University of Utah.

The process regarding my promotion to Associate Professor of Pediatrics with tenure began in the summer of 1983, before Dr. Mamzer became Chair of the Department. The vote at all levels of the University I was told, from the Department of Pediatrics up to and including the President, was unanimous, 36 - 0. I was no longer a junior faculty member and, more importantly, I was living my life-long dream. Outside the University, Bonnie and I had cultivated many friendships and we very much enjoyed living in Utah.

****

Please permit me at this point to identify the key members of my soap opera/three-ring circus cast, before I recount several sagas or stories that describe the last several years of my tenure at UU:

Drs. Kurveh, CD, and SP; and Ms. XX, an administrative assistant: Members of the UU Division of Medical Genetics in the Department of Pediatrics

Drs. Mamzer and Heartless: Chairman and Vice Chairman, respectively, of the UU Department of Pediatrics

Dr. Goniff: A faculty member in another department at the UU Medical School

UU Medical School Dean #1 (who later became VP for Health Sciences)

UU Medical School Dean #2

The Director of the Development Office of the UU Medical School

Robert (Bob) Glass, an Assistant Dean of the UU Medical School

The UU VP for Research

The President of the University

Michael and Phyllis Walton: two good friends from Salt Lake City

The Chair of the UU Retention, Promotion, and Tenure (RPT) Standards and Appeals Committee

The Chair of the UU Academic Freedom and Tenure Committee (AFTC)

Senator Orrin Hatch, R - Utah

Dr. AH: A Colorado Department of Health physician and the Director of the Mountain States Regional Genetic Services Network (MSRGSN)

The Chair of the Data and Evaluation Committee of the U.S. Council of Regional Networks (CORN) for Genetic Services

The Associate General Secretary of the American Association of University Professors (AAUP), Washington, D.C.

****

The Saga of Dr. Kurveh

11 April 1985: A memo I received from Dr. Kurveh

> *Regarding next year's salaries for the physicians in our Division, it is important to note that our Division is now on Plan B. From my discussion with our chairman, [Dr. Mamzer], your new base salary will be $45,000. Your private practice income [or PPI] supplementation will be $22,500, for a total annual salary of $67,500 beginning July 1, 1985. Your private practice income will continue to come from monies generated by the chromosome laboratory. Being on Plan B also means that our Division will pay a 2% tax on all*

> *patient income to the Dean of the School of Medicine, and an 8% tax to [Chairman Mamzer].*

Faculty members in clinical departments in the Medical School, for example, pediatrics, obstetrics and gynecology, internal medicine, psychiatry, or surgery, were usually paid from two different pots: 1) "base salary" which was usually derived from university (state) sources and/or grants and contracts from outside the university, and 2) "PPI" which was derived from seeing patients, and overseeing or working in a clinical laboratory.

The method of PPI remuneration described in this memo was inconsistent with UU guidelines. We were no longer on Plan A or Plan B - we were on "Plan M/K (Mamzer/Kurveh)." This memo was a harbinger of future renegade actions.

14 May 1985: A memo to the members of the Division of Medical Genetics from Dr. Kurveh

> *The following notes may border on rambling and redundancy, but I'll go the "memo route" and document them anyway. Without trying to preach to the converted, let me just indicate some personal thoughts about identification of our principal goals. Firstly, one of the difficulties with our "system" is that the Division of Medical Genetics and its professionals have such multifaceted goals. Delivery of service to patients "in several realms," professional and community education (again in several realms), creation and development of programs and academic scholarship (publications, scientific presentations, journal reviews, etc.) are all entwined into our jobs.*
>
> *The goals of our office staff, of course, are to support this delivery. The office is our principal overhead, i.e., time spent on administration of budgets, paying of bills, transferring of funds, help with the billing of patients, is all in addition to the secretaries' duties in helping with the above professional goals. What I mean is the job of the*

office is not merely to type up letters for patients, take patient phone calls, type manuscripts/slides and grants, but also carry out the related activities that are required for the services to exist. Thus I am making a distinction between those office duties that are primarily related to direct fulfillment of our goals and those duties that are necessary but indirect.

Two other points are important. First, a re-identification and statement of what the priorities/goals of our division are; and second, identification of where the problems are in our Division's system. I suspect that if I went to the [genetic] counselors and physicians I would probably hear a similar but slightly different statement about the ordering of our goals. Thus, I can only speak for myself. Despite the fact that teaching, research, patient care and program development are all intertwined with each other, delivery of clinical service to patients is notches above the rest on the priority list. With this in mind, I would identify that our three biggest problems in carrying out this goal of patient service to the ideal degree are as follows:

1) Getting letters out to patients and physicians in a reasonable amount of time: There are a number of variables that prevent this from flowing properly....

2) Billing: I consider that the ideal and reasonable billing process should be that the patients seen in the Tuesday morning and Thursday afternoon clinics should have completed the billing process within a week....

3) Getting things through my office: This, of course, is my problem and I am attempting to develop...a smoother way to get material in my office filed, delegated, or thrown away rather than piled up....

> *P.S. In my typical style, this* [two and a half page, single-spaced, typed] *memo has become a manuscript.*

Dr. Kurveh's rambling and redundant memos and letters, like the one above, were commonplace. Due to his lack of competence and/or interest, the clerical and billing problems only got worse as his unfinished patient charts and incomplete billing forms piled up, literally knee-deep, in his office. In addition, his job description for university-based, tenure-track physicians **[G]** was flat out wrong. What he wrote was the perfect job description for physicians either on the clinical track **[G]** or for physicians in private practice.

31 May 1985: A memo Dr. CD and I received from Dr. Kurveh

> *This memo is designed to document our agreed assignments for patient care responsibilities for the 1985/1986 academic year (July 1, 1985 to June 30, 1986). The three of us have come up with a clinical equivalent scoring system for trying to divide the clinical responsibilities of the Division of Medical Genetics on as even basis as possible. It is my understanding that we decided the following: (1) Inpatient hospital consultations will be divided evenly such that each of us will do four months per year and, (2) Responsibility for Bob's overseeing the chromosome lab will be considered equal to the responsibility of [Drs. CD and Kurveh] being in our two, half-day a week medical genetics outpatient clinics; i.e., on Tuesday mornings and Thursday afternoons.*

In fact, this was the apportionment of clinical/patient care responsibilities Drs. Kurveh, CD, and I agreed upon. My overseeing the chromosome lab (the PhD laboratory technical director I hired had left by this time), which was performing in total almost a thousand studies and other procedures a year, was considered equal to seeing patients two half-days a week.

30 September 1985: A letter I received from Dr. Kurveh

> *I felt it necessary to write you regarding the division of responsibility in the Outpatient Genetics Clinic. I hope in all of our conversations you have not interpreted my comments as taking away your contributions to genetics statewide and at our university.*
>
> *Our dilemma and I truly say "ours," is the issue of the basic division of patient care responsibility among [Dr. CD], you and me. It may seem like I am trying to funnel away some of the work that I have created for myself; however, what I am really trying to figure out is how we can re-organize this workload. The difficulty for us is that we only have [Dr. CD] and me, and two master's level genetic counselors in [the outpatient Genetics] clinic, and our workload has increased voluminously in the last three years. I know you and I differ on some of these perceptions. I am not sure that we can reach an exact compromise, but I would like to propose the following: that you attend outpatient Genetics Clinic regularly for a defined period of the year. I would propose a minimum of four months a year.*

Dr. Kurveh was the cause of the lack internal capacity and infrastructure, patient-related, problems our Division was experiencing because he loved seeing patients and he was an excellent diagnostician. Now, he was trying to renege on the agreement we had made. I had warned Dr. Kurveh at least a half dozen times that we worked in an <u>academic</u> institution, and he should stop traveling around Utah, Idaho, and Wyoming in search of more patients for Division members to see. I told him if I had wanted to spend the vast majority of my time seeing patients and overseeing a fee-for-service chromosome laboratory, which was what the lab was, I would have gone into the private practice of medical genetics years ago.

On October 8th, I sent Dr. Kurveh a letter declining his offer to see patients in the outpatient clinic for a "minimum of four months a year." The division of labor we had agreed to previously was fair and equitable, and I was already performing collaborative research with several other faculty members.

This disagreement over my not agreeing to see patients in the UU outpatient medical genetics clinic was pretty much the opening battle of what would become my long, multi-front war against incompetence and evil.

It took me almost six years of hard work to help create the Division of Medical Genetics. I kept my promise, and I expected members of the Department of Pediatrics to uphold their end of the deal - to allow me to do research for 50% of my time. In addition, because of the millions of dollars my grants and contracts to the University and the monies derived from the chromosome laboratory I was responsible for were paying the salaries and fringe benefits of almost all of the people in our Division (and also taxes to the Department of Pediatrics and the Dean of the School of Medicine), I had no intention of doing their jobs, too!

4 February 1986: A letter I received from Dr. Kurveh

> *I am concerned about funding for all the people in our division including the genetic counselors and doctors going to meetings, presenting papers, making slides for teaching and memberships in organizations. Right now the mechanism is probably not fair for our master's level genetic counselors since there is a discrepancy in generating income between them and us. In addition, I have been fairly liberal about spending my own development fund money for meetings* [that usually only he attended], *books, research, etc. For this reason, I propose we combine all the development funds in the Division of Medical Genetics into a single fund whereby it would be automatic for everyone who was going to present a paper and have an allotted amount for the year as far as other projects.*

Dr. Kurveh's development fund at this time had very little money in it because it took many months for him to submit his bills to get paid for the inpatients and outpatients he saw. There was no excuse for this. In addition, he was a free spender with what little money he had in his development fund.

In regard to the chromosome lab's development fund, I was fairly frugal spending the money in it. Unfortunately, in the end and after much discussion, Dr. Kurveh, with Chairman Mamzer's support, gained control of most of the money in the lab development account. I figured it wouldn't take long for him to spend all the money from the lab account, and I was right.

12 September 1988: A memo Dr. Kurveh sent to all of the members of the Division of Medical Genetics

> *Because of the obvious confusion surrounding the funding difficulties in our division, I felt that I would summarize some of the data [Ms. XX] and I have pulled together over the last few months. Our funding stability is much different from the Spring quarter of 1988 compared to previous years. This particular change was a finding I was aware would come, but I was not really aware of its full impact until June of this year when I realized we could not pay our taxes to our Chairman and to the Dean.*
>
> *In the last six months [Ms. XX] has been reviewing our accounts in great detail in preparation for the planning of the present fiscal year. Going through these details over the last four months I learned more about our accounts than I had during my entire tenure as Division Chief [since the end of December 1983 or almost 5 years]. I will summarize some of these facts in the next few paragraphs.*
>
> *First I think it is important to note that the salary budget of our division for the 1987/1988 fiscal year is $869,000. Given the increases for fiscal year 1988/1989 and our divisional expenses (phone bills, office supplies, laboratory supplies), the budget of our division is about one million*

> dollars. The fact that we were managing that degree of funds was certainly astounding to me until we looked into it from this point of view. [The fact that it took several years for Dr. Kurveh to appreciate the size of the Division's budget was, and still is, beyond my comprehension.]
>
> One of the financial facts that did not become clear to me until our recent review was the fact that in fiscal year 1987/1988 (compared to fiscal year 1986/1987), our division had to come up with an additional $85,000 in salary money compared to the previous year. This amount of money was something that we were all aware of. I certainly was not predicting its impact to have such a far reaching effect in June as to make it that we could not pay our departmental and dean's taxes or in July that we had to actually wait for all the income from the lab and clinic to completely pay bills on our accounts.

By now, all the money Dr. Kurveh had appropriated from the chromosome laboratory's development account was spent, and the Division was in severe financial distress. When the U.S. Supreme Court in 1982 declared the practice of medicine to be a business rather than a profession, it certainly wasn't thinking of Dr. Kurveh, because he didn't know anything about running a business or being an administrator. His division chief's job, however, was not in jeopardy as long as he continued to be a sycophant for Chairman Mamzer. In addition, he had to quickly find some scapegoats to relieve the Division's financial crisis.

12 October 1988: A letter I received from Dr. Kurveh

> It turns out that last Monday at the [weekly] Division of Medical Genetics meeting you spoke [protested] too soon in regards to [Dr. SP] being the only one affected financially and the rest of us not. In my two meetings with [Chairman Mamzer], it is clear that there are no other options. I have no choice but to cut our private practice income, and I have done such. I have

not cut [Dr. SP] further since she has already had her [base] salary cut by $8,000 this year. I have cut my private practice income check, yours, and Dr. CD's by 30 percent. [The 30% cut in my PPI represented about a 10-15% cut in my total annual income.]

Obviously this affects me personally and I regret it tremendously. It has had an effect on my own personal life and finances that are clear cut. Certainly, if our division's financial situation improves, then we can reinstate our appropriate salary amounts. Unfortunately, there is just no money that the Department has to assist in such a situation. I still feel that it was my planning and risk-taking that is responsible for us not having enough money to pay [Dr. SP] at the junior professor [a type of faculty position known only to Dr. Kurveh], *so anything you say has already been thought by me.*

Dr. SP had spent two years working as a post-doctoral fellow in our Division; that is, the same kind of position I had held at Yale from 1974 to 1977. She was then offered a limited-term instructor position in the Division for $50,000 a year, beginning 1 July 1988. I felt very sorry Dr. Kurveh reneged on his promise to Dr. SP, a top-notch young faculty member working at her first job. However, there was nothing I could do at the time other than offer her my condolences - which I did.

In June, 1989, Dr. SP left the UU in disgust and our Division was back where it started - too many patients and not enough physicians to see all of them. In addition, what Drs. Kurveh, Mamzer, and CD didn't know was my friends in the Department of Pediatrics gave me written intelligence information that showed Dr. CD never took a cut in her PPI, and Dr. Kurveh's PPI pay cut was more than offset by an end-of-the-year (December, 1988) bonus he received. In the end, only Dr. SP and I took pay cuts, and these were clearly in violation of University regulations. Under Dr. Kurveh's leadership, renegade behavior became the norm in our Division, with the full support of Chairman Mamzer, of course.

Finally, Dr. Kurveh said at the end of his letter, above, "so anything you say has already been thought by me." What I wanted to ask him at that time was, "How do you live with yourself; how can you look at yourself in a mirror; and how can you sleep at night knowing the injustices you have heaped on others? Have you no decency or sense of shame?" I didn't ask him these questions because it would have been fruitless, and because in many ways he was inept, incompetent, and dishonest - and I wouldn't have believed anything he said anyway.

16 November 1989: A letter I sent to Dr. Kurveh

> *It has just come to my attention that you misappropriated funds from my "Reproductive Medicine Symposium" account. As you know, that account is in my name and cannot be used without my approval. The funds are, in part, from a grant I received from the Herbert I. and Elsa B. Michael Foundation....Because of the seriousness of misappropriating funds from this account, I request that you replace the funds immediately so that they can be used in accordance with the terms of the grant and my wishes.*
> *cc: Director of Research Accounting, UU*

1 December 1989: A letter I received from the Director of the Department of Research Accounting, UU

> *As you requested, we have reviewed transactions recorded in the "Reproductive Medicine Symposium" account. The signature card on file for the symposium account in the Accounts Payable Department lists your name and [Ms. XX], Administrative Assistant in the Division of Medical Genetics, as authorized signatures. The card was dated January 28, 1985. Although this review was limited, we believe it indicates the following issues that you and the administration of your department need to resolve.*

The Reproductive Medicine Symposium account was established in 1982 with a $15,000 grant to you from the Herbert I. and Elsa B. Michael Foundation. The grant was restricted for use in support of the first symposium held in 1982. We understand that the $15,000 was expanded for this purpose, and that the balance of $8,200 remaining on July 1, 1989 was comprised of additions funds generated by additional symposia, along with interest earnings. For ongoing programs, such funds normally carry the same restrictions as an original grant or gift.

It was improper for [Dr. Kurveh] to charge his travel expenditures to a restricted account like the Reproductive Medicine Symposium account, and then deposit reimbursements for those same expenditures to an unrestricted account controlled by [Dr. Kurveh]. The applicable amounts should be returned to the Reproductive Medicine Symposium account.

In addition, it appears that a $528 travel reimbursement from [Dr. Kurveh] due to the University since March, 1989, has not been received. This should be pursued.
cc: [Dr. Kurveh]

6 December 1989: A letter from [**Dr. Kurveh's wife**] sent to Ms. XX

This letter is meant to document the circumstances surrounding the $528 reimbursement for the airfare for my husband's trip to Riverside, CA in March, 1989. Since I am in charge of our family's finances, I deposited this money directly into our personal account. It was not until a few weeks later I was informed by my husband that $528 of the reimbursement was owed back to the Division of Medical Genetics for airfare. It has been known since that time that the money was owed, but paying it back has

> been merely overlooked. Enclosed please find a check in the amount of $528 to correct this oversight.

It took Dr. Kurveh and his wife nine months to reimburse my Reproductive Medicine Symposium account the $528 it was owed - almost the same amount of time it had taken Dr. Kurveh to bill his patients. Do you think Dr. Kurvah and his wife would have reimbursed the account if I didn't discover their "oversight?" Ha!

Shortly thereafter, Dr. Kurveh stole the remaining money that was left in this restricted account and placed it in his development account, with the blessing of Chairman Mamzer,of course.

1 March 1990: A memo I received from the Director of Research Accounting, UU

> Your [Utah] Department of Health account is over-committed in the amount of $31,038 and has been frozen. No new encumbrances or unobligated expenditures can be processed until the account changes to a positive uncommitted balance, at which time your account will be reactivated.

6 March 1990: The letter I sent to the **VP for Research** in response to the March 1st memo I received from the Director of Research Accounting

> This letter will serve as a follow-up to our phone conversation. As I noted to you, $33,091 from my State Department of Health contract has been used by the Division of Medical Genetics for purposes other than the terms of the contract. This extraordinary use of funds resulted in a $31,038 deficit, making it impossible for the contract to be executed as intended. I draw this to your attention because as principal investigator of this contract I have both a fiduciary and moral responsibility to ensure that it is properly executed. The monies were appropriated without my signature or consent. I respectfully request

> *that you exercise your office to investigate this matter and, if my information is correct, see that the problem is rectified immediately.*

I never received a response to this letter. It was stone-walled because much of the hierarchy/administration/mafia at UU, and especially in its Medical School, was as morally bankrupt (call it, *honor among thieves*) as the Division of Medical Genetics was financially bankrupt.

****

> *That which does not kill us makes us stronger.*
> Friedrich Nietzsche (1844 - 1900)

In the Darwinian, survival of the fittest environment that was the Department of Pediatrics, only the strong and the resilient survived. With help from my family and friends, I survived and flourished academically-speaking, for the first several years of Dr. Mamzer's chairmanship. This made me both happy and proud. On the other hand, I presume it caused Drs. Mamzer, Kurvah, and CD to dislike me even more.

11 March 1987: A letter written by a Professor in the Department of Pediatrics to the Chair of the <u>Department of Anatomy</u>, UU Medical School - this was one of several letters written to the Anatomy Department Chair

> *This letter is in support of a secondary appointment for Dr. Robert Fineman as a Research Associate Professor in the Department of Anatomy.*
>
> *I have known Bob since 1977 when he arrived as an Assistant Professor and Chief of the Division of Medical Genetics. Bob's initial efforts were in establishing broad-based genetic services, not only at the University of Utah Medical Center, but also throughout our state....*

> *More relevant to his research appointment in Anatomy are his investigative activities. He has thirty-seven publications, and is first author on fourteen of them. In an additional, six or seven publications, a post-doctoral fellow or medical student is first author and Bob is senior author....*
>
> *In summary, Bob is clearly an established independent investigator who has received national recognition for his work in medical genetics and cytogenetics. Importantly, with regard to his appointment in Anatomy, he has an established collaborative relationship in research projects with several members of the department. I can, therefore, recommend him for appointment as Research Associate Professor of Anatomy with highest enthusiasm.*

26 June 1987: A letter I received from the President of the University

> *I am pleased to advise you that the University's Institutional Council has given formal approval for your secondary appointment to the non-tenure track faculty of the University of Utah as Research Associate Professor of Anatomy, effective immediately....*

1 February 1988: A letter I received from the President of the Society for Pediatric Research

> *I am pleased to inform you of your election to membership in the Society for Pediatric Research. The Council and members welcome your active participation. Your election to this Society represents peer recognition of your research achievements and independence as an investigator.*

****

The Saga of Dr. Mamzer

Power [narcissism, hypomania, and/or testosterone poisoning] *confuses itself with virtue and tends also to take itself for omnipotence.*
J. William Fulbright (1905 - 1995)

4 April 1988: A letter I received from Chairman Mamzer

I am writing to summarize our conversation of earlier today. I am delighted with how well the birth defects registry project is moving....We are both very hopeful that it will come to fruition within the next three months.

You outlined for me very clearly how you spend your time and the reasons for which you think a pay raise is justified as of July 1, 1988 [several days prior to this April 4th meeting Dr. Kurveh told me Chairman Mamzer approved raises for everyone in the Division of Medical Genetics except me - ergo, one of my reasons for requesting this meeting]. *You outlined your 35% time commitment to the chromosome lab and the approximately $230,000 in new grants and contracts which you have obtained as principal investigator. You have also arranged to have an additional 23% of your salary paid by the State Department of Health in July, and you have arranged for 10% of your salary to come from the Mountain States Regional Genetic Services Network. You are also hopeful that another large ($360,000) federally funded grant may become available in the near future.*

I explained to you the reason why you will not get a pay raise in July of 1988. Your proposed salary puts you in the 20th percentile nationally for Associate Professors. I believe your performance as an Associate Professor falls well below the

20th percentile performance. Your performance as an Associate Professor in the Department would put you in the lowest 5%.

We discussed in some detail the discordancy of your view of people and the world and that of others. That discordancy also has a behavioral correlate in your case, which leads to gross insubordination within the Division and Department. You are a cause of general discontent in the Division because of your refusal to follow directives agreed upon by the Division and requested by the Division Chief. You consume a huge amount of time and create immense agitation within the Division, and derivatively within the Department. Your particular insubordination with reference to the Department Chairman is well known, and we again reviewed the situation. Your disruptive interactional behavior spreads throughout the School, and is, at a minimum, exasperating to your colleagues and supervisors.

I believe your rankings as a clinician are much below average, and your ranking as a clinical teacher to house staff and as a clinical consultant to your peers is also much below average….

Any future salary increase will depend on how you have addressed the problems of your behavior in our complex environment. I expect you to do what your Division Chief asks, and to do it in a way that assists the Division rather than disrupts it. You need to be supportive of the career development of other faculty in the Division. I expect less institutionally disruptive behavior and less institutional insubordinate behavior. Finally, I expect a better translation of your passion for patients and their health as you teach in a clinical environment.

Of the hundreds of documents I collected during my time at UU, this letter is the one I treasure the most because it describes a true "a-ha!" event in my career.

This letter was the perfect vehicle for my advisors and me to understand Dr. Mamzer. If I were such a poor clinician/diagnostician, why were Drs. Kurveh and Mamzer busting their chops to get me to see more patients in clinic? If I were a poor clinician one would think they would do everything possible to keep me from seeing patients, so that I wouldn't harm any of them. Also, if the medical students and the pediatric interns and residents thought I was such a poor teacher, why did they vote several months later as a group in favor of my promotion to Full Professor? And if my peers, the physicians in Utah, thought I was a lousy consultant, why did they vote several months later as a group in favor of my promotion to Full Professor?

What was even more interesting about his letter was, during our conversation, we never discussed "discordancy" or "insubordination," or anything about my abilities as a clinician, teacher, or consultant. In addition, Dr. Mamzer left out of the letter any mention of the discussion we had about Dr. Goniff - whom you will get to know a lot better later in this chapter in the section entitled: **The Saga of Dr. Goniff.**

Amazingly, only about half of Chairman Mamzer's letter was based in reality. The rest was pure fabrication on his part. His remark, "You need to be supportive of the career development of other faculty in the Division," was a ruse. My grants, contracts, and monies from the chromosome laboratory were already paying for most of the salaries and fringe benefits of the members in the Division of Medical Genetics, including Drs. Kurveh, CD, and SP. In fact, what Chairman Mamzer wanted was for me to see more patients and for the chromosome laboratory to become a commercial entity, rather than an academically-oriented laboratory like the one at Yale, in order to generate larger sums of unrestricted tax money for both the Department and the Medical School. Several years after I left the UU on 1 January 1991, the chromosome laboratory I had directed was performing about 10,000 studies a year.

It should also be noted that all of my salary, fringe benefits, travel expenses, telephone, books, postage, and office supplies, during the last several years of my time at UU were paid for by outside of the University sources. In essence, I was a totally free ride for the UU.

Finally, after I read his letter, above, it was clear to me what I had to do two months later. I asked to be promoted to Full Professor! I did this not only to give Chairman Mamzer more rope to hang himself, but also because it is good to have a sense of humor in situations like this.

****

Prior to and after our conversation on 4 April 1988, Chairman Mamzer and I had many unusual conversations and, at this point, I would like to tell you about four of them.

In January 1984, several months after he arrived at UU, I called him one evening at his home to seek some advice about a grant application I was writing. He had been out of town for several days and I was leaving town the next day for several days. I wanted to write part of the grant application while I was out of town. I started the conversation with an apology for calling him at home. Before I had a chance to ask my question he said aggressively, "How dare you call me at my home! Who the hell do you think you are! Don't you ever call me at home!" I tried to apologize again but he wouldn't have it. He told me to call him at his office and he hung up.

Another time, when I called Dr. Mamzer at his office and, unfortunately, I can't remember the question I asked him, he responded by shouting in the phone seven times in rapid succession, "It's your fault!" That was all he said. I told him I only wished I had taped the conversation, and I said goodbye. Shortly thereafter, I started secretly taping my conversations with the Drs. Mamzer, Kurveh, and several others. It was perfectly legal in Utah to do this.

In the spring of 1984, I had a discussion with Chairman Mamzer in his office. During the conversation I told him the master's level genetic counselors in the Division of Medical Genetics were significantly underpaid according to their level of expertise and experience. Dr. Mamzer said I should "get rid of them and hire recent graduates from master's level genetic counseling programs," implying that we could save money that way. I objected, saying we should be loyal to the good people we have. He then said, "Bob, you should leave Utah because you don't do anything around here." His refusal to properly address this genetic counselor issue caused significant morale problems in the Division.

In April 1987, I met with Dr. Mamzer and told him the chromosome laboratory at the Children's Hospital had made recently several serious errors in the interpretation of its chromosome studies. I asked Dr. Mamzer for his advice because I was the medical geneticist seeing, in the Children's Hospital Birth Defects Clinic, patients who had had their chromosome studies performed at Children's. Obviously, I wanted to avoid any medical-legal issues and political fall-out that could include me, the University, and Children's Hospital. Dr. Mamzer responded, "Bob, you are always looking for trouble. I have a lot of headaches and you are just another one of them."

One of reasons I enjoyed being a member of Rotary is because Rotarians around the world promote high ethical standards in the workplace through use of the "Four-Way Test" which asks the following questions: "Of the things we think, say or do: Is it the truth? Is it fair to all concerned? Will it build goodwill and better friendships? Will it be beneficial to all concerned?" Chairman Mamzer's actions and comments, and the goal of the Four-Way Test, represent two ends of a spectrum - good versus evil. Among other things, he was a highly abusive bully who lacked self-control (see Appendix C, Document 1, for yet another example of his alleged bad behavior).

I need to tell you at this point about **Michael and Phyllis Walton**, two very talented, ethical, and prominent business people/friends of mine from Salt Lake City. The Waltons helped me immensely during my struggles at the UU. In fact, they co-authored many of the letters, memos, and other documents I submitted to individuals and committees there, and elsewhere.

Michael received a PhD in the History of Science from the University of Chicago and he attended the UU Law School. Phyllis graduated from the UU Law School. Without their advice and support I never could have engaged Chairman Mamzer because virtually everything regarding my struggles had to be framed in legal terms and concepts. Ethics and morality, obviously, counted for very little at the UU Medical School. I had no training in law, and the amount of time it took for the Waltons and me to defeat Dr. Mamzer was considerable. For example, in 1990, toward the end of my struggles at UU, the following happened. Nathan Walton, the then four-year-old son of the Michael and Phyllis, was in a car being driven by this older sister, Josie:

> Nathan: "Look, Josie, there is our house."
> Josie: "Yes, Nathan, and who lives there? "
> Nathan: "Mommy, Daddy, Josie, Nathan, and Bob Fineman."

As for the advice I received from the Waltons regarding Drs. Mamzer, Kurveh, and their ilk, Michael summed it up very well and very succinctly: "Bob, they wouldn't piss on you if you were on fire!" He also said, "You are the only person I know who would file a grievance against your Chairman and your Division Chief and then, shortly thereafter, ask to be promoted to Full Professor."

In fact, when I did these things I knew exactly what I was doing and what Chairman Mamzer would do in response. The 4 December 1989 report (Appendix C, Document 2) from the **Chair of the University of Utah Retention, Promotion, and Tenure (RPT) Standards and Appeals Committee to the President of the University** regarding my appeal for promotion to Full Professor confirmed my prediction.

****

Since "know your enemy" was and still is a cardinal rule of fighting a war, I achieved this convention in large part by having good advisors, like the Waltons, Don Summers, and others, and by creating an elaborate intelligence gathering system throughout the Medical School and the University. The system, whose members thought they were acting alone and actually never knew a system existed, provided me with many useful documents and other relevant information.

Using this information, guidance from my advisors, and some intuition, I determined the best way to confront Dr. Mamzer, who fought according to the old adage, "If brute force hasn't worked, I must not be using enough of it," was to be patient and use the "rope-a-dope" strategy perfected by Muhammad Ali (b.1942).

At first, because I was naïve, I had a difficult time figuring out what kind of person Dr. Mamzer was. When it became obvious to me what kind of person he was, I couldn't believe it. Then I thought to myself, others like the Dean of the Medical School, the V.P. for Health Science, and the President of the University, will come to my rescue. Wrong! In the end, I was forced to work with my friends, including my intelligence gatherers both inside and outside the University, in secret, to improve transparency at the Medical School and to protect services to thousands of children and their families living in Utah, Idaho, and Wyoming.

Dr. Mamzer had predicted at one point that I had the potential to be "fatal" to him and his cronies. In the end, that was, in fact, the only correct thing he ever said about me! While no one died, I was able to expose them for who they really were.

16 July 1990: "University Intensive Care Unit Divided on Patient Limit - Doctors, Nurses Dispute Care-Giving Capabilities" *The Salt Lake Tribune* - a Salt Lake City daily newspaper

> *An administrative decision to limit the number of patients in the University of Utah's nationally recognized intensive care unit (ICU) has top physicians and nurses at odds....*

This newspaper article marked the beginning of the end of Dr. Mamzer's painful and costly chairmanship at UU. If the leadership of the Medical School and University had a moral compass, Dr. Mamzer would have been relieved of his position as Chairman much sooner, and a tremendous amount of needless suffering by many individuals would have been prevented.

27 September 1990: "University Hospital Audit to Probe Antitrust Allegations" *The Salt Lake Tribune*

> *An audit will be performed at the University of Utah School of Medicine and Hospital to ensure that the institutions are not violating federal and state antitrust laws, a University attorney said Wednesday. The chairman of the Department of Pediatrics, [Dr. Mamzer], is also medical director at the Children's Hospital, which is owned and operated by Intermountain Health Care Inc., the largest health-care conglomerate in the state. The Children's Hospital moved close to the University campus in April. Some University Intensive Care Unit personnel, including doctors, contend it is a conflict of interest for [Dr. Mamzer] to maintain both positions. The Attorney General's investigation apparently stems from the proposed agreement between the University Hospital and Children's that is similar to an agreement [that was] blocked in 1985....*
>
> **[In 1985], the Attorney General's Office reviewed the proposal at the request of the State Board of Regents and determined the proposed allocation of services was illegal under the federal Sherman Act and the Utah Antitrust Act. "If the parties were to enter into agreements as drafted, both the University of Utah and Children's Hospital would be vulnerable to antitrust liability," the formal opinion by the Attorney General's Office stated [emphasis mine].**

11 October 1990: "Actions of Doctors Group Targeted" *The Deseret News* - a Salt Lake City daily newspaper

> *Blue Cross and Blue Shield is questioning whether a group of physicians may have violated antitrust laws by forming a private corporation that attempts to negotiate fixed physician fees. Meeting those fees could force Blue Cross,*

Utah's largest insurance company, to raise its rates.

A senior vice president and general counsel for Blue Cross Blue Shield of Utah said for legal reasons he could not comment on the insurance company's antitrust concerns. But he did say Blue Cross has been approached by Pediatric Faculty Physicians (PFP) Inc. about negotiating a contract with the insurance company based on a fee schedule the physician corporation drew up.

PFP Inc. comprises "every full-time faculty member of the Department of Pediatrics who is a medical doctor," said [Dr. Mamzer], who is chairman of the board of PFP Inc., chairman of the U Department of Pediatrics and medical director of Children's Hospital. [Dr. Mamzer was wrong. I never joined PFP Inc. because I knew it was a sophisticated way to steal from the patients.]

[Dr. Mamzer] said PFP Inc. does not violate antitrust law. "To the contrary," he said, "PFP was formed at the request of insurance companies who would rather deal with groups of doctors than individual doctors. He estimated between 80 and 85 doctors belong to PFP Inc....

The Attorney General's Office is investigating the U Hospital and Children's Hospital for possible antitrust violations, according to several physicians at both hospitals. The investigation, they said, focuses on the inpatient care units of both hospitals, plus the U Department of Pediatrics and its intensive care unit - programs in which members of PFP Inc. are key players.

"PFP Inc. appears to be a monopolist in this region for critical pediatric care. Moreover, PFP appears to be attempting to exercise this monopoly power to exact higher fees from Blue

> Cross for services that PFP alone offers the community," [he] said in a memo.

1 November 1990: "Memo from Physician Group Urges Members to Charge Maximum Fee" *The Deseret News*

> The president of a private corporation of physicians sent a memo to the doctors in his company urging them to bill insurance companies at or above negotiated rates in order to get the maximum dollars from the insurance companies. The corporation, Pediatric Faculty Physicians Inc., says it is the only one providing critical subspecialty care in the region.
>
> Pediatric Faculty Physicians, a group of University of Utah doctors, has negotiated fixed fees with several insurance companies. The doctors, who also practice at Children's Hospital, an Intermountain Health Care facility, were told not to negotiate individual contracts with at least one insurance company, documents show. That instruction has since been rescinded, however.
>
> Two insurance companies claim PFP's fee demands are substantially higher than current fee levels paid by those companies and would increase health-care costs for their customers.
>
> In a March 26, 1990, memo obtained by The News, [Dr. Heartless], PFP president, instructed doctors in his corporation to "please be sure your billed charges are equal to or above the highest fee listed for those codes you presently bill for." [Dr. Heartless] is vice chairman of the Department of Pediatrics and medical director of the Children's Hospital intensive care unit.
>
> "These fees represent the maximum reimbursement. If you charge less, they will pay less," he wrote. [Dr. Heartless] told The News Thursday, "You wouldn't want to bill below the

> *rate we agreed upon with insurance companies or the physicians wouldn't be getting the maximum benefit."*

Dr. Heartless was a long-time friend of Dr. Mamzer. He became a member of the Department shortly after Dr. Mamzer became Chairman. Drs. Mamzer and Heartless shared many of the same personality traits.

8 November 1990: "University Doctors Ask [UU President] to Counter Publicity - Doctors Say Chairman, Department Have Gotten Bum Rap in Controversy Over University - Children's Hospital Ties" *The Deseret News*

> *Senior faculty members of the University of Utah's Department of Pediatrics want the U President to counter a tide of negative publicity they say their department and its chairman don't deserve. In a November 6 letter to [the President], twelve faculty members said, "We are distressed by the negative publicity that both our department and its chairman have received." The faculty members said the media had made little attempt to uncover the facts, rather than basing their reports on opinions of individuals.*
>
> *The News attempts to obtain comments from faculty members over the various controversies have been thwarted for the most part by the faculty's refusal to comment on the record. They have also refused to comment, saying they are doing so on the advice of their attorneys.*
>
> *The letter also criticized the university and medical school administrations, saying the two administrations have "done little to counter the adverse publicity. We feel that it is time for the administration of the university and medical school to address the allegations in a reasonable, unemotional, accurate and factual manner."*

It never ceased to amaze me that each time Chairman Mamzer caused a problem, he blamed someone else, and the weak, unethical administration at the Medical School and University agreed with and supported him. This continued for more than a decade.

13 November 1990: "Probe of Utah Health Industry Reflects National Antitrust Drive" *The Deseret News*

> The Utah Attorney General's investigation into Utah's health-care industry appears to be in concert with a nationwide drive by the U.S. Department of Justice and the Federal Trade Commission to crack down on health-care organizations across the country violating the Sherman Antitrust Act. The October 27, 1990, issue of American Medical News (AMN) detailed the federal government's aggressive investigations of price-fixing, boycotting and other antitrust activities in the health-care industry. Three Arizona dentists were convicted in October for price-fixing, the AMN reported.
>
> Still, one University of Utah doctor who is part of the investigation accused the Attorney General's Office of conducting a "witch hunt." "There aren't very many things in life that I describe as good and evil. But that office is evil," he said. "What they are going to tear apart by their behavior with reference to the University of Utah is unconscionable. The Attorney General is on a witch hunt that someone in the press ought to investigate. That office is in total disarray," said the physician, who wished to remain anonymous. [One can only surmise the anonymous doctor was highly placed in the Department of Pediatrics.]
>
> The comment didn't sit well with the Attorney General's Office. "No one to our knowledge has ever accused this administration of a witch hunt before. We certainly aren't doing that," said the

Chief Deputy Attorney General. "As to the particulars of the investigation, it is our policy not to comment on them. I can tell you we are looking into healthcare in Utah generally for antitrust concerns."

17 November 1990: "University Hospital Lays Blame - On Media" *The Deseret News*

University of Utah Medical Center officials blame the media and a few disgruntled employees for problems that have surfaced between the U Hospital and Children's Hospital and have sent a memo to U employees threatening them with discipline if they speak anonymously to the media....

The investigation, which has been confirmed by several physicians as well as outside counsel hired to represent U Hospital, has been the focus of media attention for several weeks.

28 November 1990: "University Hires Lawyers in Hospital Probe" *The Deseret News*

The University of Utah has hired a team of lawyers from three law firms [located in Washington, D.C., and at taxpayers' and patients' expense] *to represent the University, a private physicians' corporation* [Pediatric Faculty Physicians Inc.], *and a university doctor [Dr. Mamzer] during an antitrust investigation conducted by the Utah Attorney General's Office....*

30 November 1990: "FHP Says Alliance is Costing It Money" *The Deseret News*

Family Health Plan (FHP), Utah's [first and] *largest health-maintenance organization* [HMO], *has discovered that the close relationship between Children's Hospital and the University of*

> Utah Hospital is costing the company a lot of money - costs FHP will pass on to its members next summer. FHP officials said they won't be able to attach a dollar figure to the cost increase for two or three months. The company has 124,000 members in Utah.
>
> FHP pays approximately 36 percent more for one day of care at Children's Hospital than a day of care at the University Hospital, said an FHP regional vice president....

9 December 1990: "Cooperation Between Hospitals Stirs Up Debate" *The Deseret News*

> Seeing an explosion in the state's health-care costs, a governor's task force concluded that the Utah Attorney General should investigate complaints of abuse by health-care providers that dominate Utah's market. The Governor's Task Force on Health Care Costs specifically suggested the state conduct a "full investigation of possible anti-competitive activities and violations of the public trust" by health-care providers.
>
> The investigation centers on the inpatient and intensive care units of the University of Utah and Children's Hospitals and a private corporation formed by physicians associated with both the U and Children's. "Competition between the only two hospitals which provide in-depth care to children is necessary in order to keep costs low and ensure the best possible care," said the chief deputy attorney general.
>
> But the President and Chief Executive Officer of Intermountain Health Care and U.S. Senator Orrin Hatch, R-Utah, aren't convinced cooperation between hospitals is bad. "Antitrust law is not biblical in origin, as your more liberal lawyers would have you believe," Hatch said.

"Frankly there is every reason in a society that is reeling from (medical) costs to try to allow cooperative efforts among various groups to try to keep costs down. Under certain aspects of antitrust laws, you are unable to do that these days - at least in accordance with some of the interpretations."

The chief deputy attorney general counters that antitrust law may not be biblical, but "it isn't written on toilet paper, either."

With all due respect to Senator Hatch, in this instance he was wrong. He was either misinformed, naïve, or he had another agenda in mind. There was no attempt to save patients and families money in the UU Hospital/Children's Hospital collaborative effort. In fact, the goal was the exact opposite. Doctors, hospital administrators, and others, who thought they had a license to steal, wanted to charge more. This was clear from the very beginning.

****

On 1 January 1991, the glib Dr. Mamzer began a one-year sabbatical to work with Senator Hatch. According to the *American Academy of Medicine News* (May 1991), Dr. Mamzer along with several other physicians worked in Washington, D.C., to "serve as health policy analysts and advisors to U.S. senators."

"Physicians are able to get their message across because of the commitment, time and caring they bring to the political arena," [Dr. Mamzer] said. "All they need to do is act on the things they care about [for many physicians, this meant/means money, control, power, and autonomy]. If they do this in concert, they can be more effective [for example, Pediatric Faculty Physicians Inc.]."

****

5 May 1991: "Doctors Take 'Interest' in Overdue Bills" *The Deseret News*

> *You think interest on your credit cards is steep? Take a second look at your doctor bills. Some area physicians are using high interest rates and aggressive collection tactics to battle a mounting deluge of delinquent accounts they say are caused by a national recession and receding employee benefits. In short, patients responsible for a large portion of their bills are taking longer to pay them - and doctors are tired of waiting.*
>
> *A group of University of Utah physicians recently decided to raise the interest on delinquent accounts to 18 percent and turn accounts over for collection after 90 days instead of the former 120 days. The physicians also practice at Children's Hospital but will bill separately from either hospital [using the bill collecting services of Pediatric Faculty Physicians Inc.].*
>
> *"I think it is just good business," said [Dr. Heartless], acting co-chair of the U's Department of Pediatrics [while Dr. Mamzer was on sabbatical] and acting medical director at Children's Hospital. He equates medical care with credit-card use. If a consumer builds up a high credit card balance, he pays a high interest rate. If he racks up a high medical bill, he pays high interest. "Unfortunately medical care is a luxury. It shouldn't be in this country, but it is," [Dr. Heartless] said.*

Dr. Heartless' go for the jugular-type remarks represent, in large part, what has been and continues to be an overwhelming evil regarding doctorhood, the medical profession, and healthcare in the U.S. They also help define, in part, the meaning of the medical-industrial complex: *Motive* - a lust for money; *Opportunity* - an enormous trough of money that was/is readily accessible; and *Means* - for physicians, a medical license.

In other words:

> *It's not personal, it's strictly business.*
> Michael Corleone, The Godfather (1972)

The families of the hospitalized children who had, for example, cancer, severe birth defects and genetic disorders, infections, trauma, asthma, and diabetes, weren't buying new cars, jewelry, or taking expensive vacations - but you can be sure a lot of the doctors were. Children were suffering terribly, and Drs. Mamzer, Heartless, and the others who participated in PFP Inc. were heaping more misery on the patients and their families.

As I mentioned before, I never joined Pediatric Faculty Physicians Inc. I threw into trash cans at least five or six letters asking me to join PFP Inc. because the organization stood for the exact opposite of what I believed physicians should be. PFP Inc. also brought back painful memories of my father's suffering and premature death. Dr. Heartless' actions and remarks were not only callous in my opinion, they were despicable!

Comments like Dr. Heartless' were to me like to waving a red cape in front of a Tauran, which is my astrological sign, if you have not already guessed. When someone waves a red cape in front of a bull, the bull charges in most cases. I had a different plan. I became a *Deep Throat* in 1990, and until now only a few people have known this.

My friends at UU had given me all sorts of letters and other documents about what was going on in the Department of Pediatrics and PFP Inc., and I passed them on to a local newspaper reporter and the Utah Attorney General's Office. I knew hundreds if not thousands of families of sick and/or dying children would suffer additional financial hardships if I didn't do what I did. I also knew that my actions were consistent with the Boy Scout Oath and Law, and the Oath I took when I graduated from medical school.

20 June 1991: "Hospital Probe in Federal Hands" *The Deseret News*

> The U.S. Department of Justice has taken over the state's antitrust investigation into the University of Utah Hospital, Children's Hospital and a group of physicians who work at both hospitals. "The problems at the U - and with overall health-care in Utah - are so large that with our meager resources we would only be scratching the surface years from now," said the Attorney General for Utah. "So we asked for help." ....
>
> Unlike the Attorney General's Office, the Department of Justice has the power to criminally prosecute antitrust violations. Sources also told The News that pressure from the U and higher education officials to drop the investigation prompted the Attorney General to turn it over to the federal government.

****

> For there are two things a prince [like Dr. Mamzer] *has to fear - the one, attempts against him by his own subjects; and the other, attacks from without by powerful foreigners* [like the U.S. Department of Justice].
> Niccolo Machiavelli (1469 - 1527)

****

28 December 1991: "U Wants State to Pay Legal Fees of Probe" *The Salt Lake Tribune*

> The University of Utah wants the state to pay attorneys to defend the school in a federal antitrust investigation into Utah's health-care industry. In fact, university officials are expected to request a special legislative appropriation for

> *legal fees that some predict will exceed $10 million....*

Apparently, there wasn't enough money to steal just from the patients. Now, the doctors were going to steal from the taxpayers, too.

6 March 1992: "U Hospital - Intermountain Health Care Probes Triggers Dissolution of Pediatric Faculty Physicians Inc." *The Salt Lake Tribune*

> *Pediatric Faculty Physicians (PFP) Inc., a corporation of about 90 doctors in the U Department of Pediatrics - many of whom practice at Children's Hospital - has been dissolved....*
>
> *While it has been widely known that PFP is the target of a federal investigation, the school's attorney said the ongoing antitrust probe was not the primary reason for its termination.*

A formal announcement by the U.S. Department of Justice, dated 14 March 1994, stated, "The...decree further provides that the University will take whatever steps are necessary to prevent the reformation of Pediatric Faculty Physicians Inc., which was involved in negotiating physicians' fees, or the creation of any similar organization."

Shortly after this announcement, Chairman Mamzer left the UU to become the Dean of another medical school. I presume the UU's hierarchy finally realized how negatively he had impacted its program. So, they probably gave him glowing references and kicked him upstairs, to another university. He was forced to resign from his next position less than two years later. Apparently, the other university was more astute in its evaluation of Dr. Mamzer.

****

In the March 1992 edition of the *Atlantic Monthly* magazine, a former editor of the *New England Journal of Medicine*, Dr. Arnold Relman, said he fears his profession has lost its ethical way.

Doctors, he argued, were not and should not be businessmen, and yet financial and technological pressures were forcing more and more physicians to act like businessmen with deleterious consequences for patients and society as a whole.

With all due respect to Dr. Relman, the truth is many doctors in the U.S. have acted like greedy businessmen for more than 200 years:

> *It is not because the truth is too difficult to see that we [doctors] make mistakes.... We make mistakes because the easiest and most comfortable course for us is to seek insight where it accords with our emotions - especially selfish ones.*
> Alexander Solzhenitsyn (1918 - 2008)

****

1 July 1993: "Price-Fixing Suit Still Alive Against Two Utah Hospitals" *The Salt Lake Tribune*

> *Jubilant announcements touting the end of an antitrust investigation against University Hospital and Primary Children's Medical Center are dead wrong, says a top Justice Department official.*
>
> *Antitrust Division Chief Anne K. Bingaman denied reports that Justice staffers recommended no criminal charges be brought in the 2 ½ -year-old investigation....It is possible that criminal charges could be recommended in one or both matters....*
>
> *"No one has reached any conclusions of any kind," Ms. Bingaman insisted....*
>
> *News of the continuing criminal probe flew in the face of announcements by the University of Utah, Utah Atty. Gen. Jan Graham and Intermountain Health Care that Justice staffers recommended no criminal charges against any targeted individuals or organizations.*

> *Most Utah news organizations reported the erroneous information, based on state officials' statements. Justice Department officials refused comment on the case until prodded by U.S. Sen. Orrin Hatch, ranking Republican on the Judiciary Committee....*
>
> *Mr. Hatch called the no-indictment recommendation on some allegations "a step in the right direction. On the other hand, I want the whole thing resolved."*
>
> *Taxpayers may not know the details of the investigation. But they are taking the brunt of it.*
>
> *The drawn-out probe has eaten up an estimated $10 million in legal expenses in tax money and patient bills....*

Chairman Mamzer had spent a sabbatical year in 1991 in Washington, D.C., with Senator Hatch. One of the lessons that should have been learned from that sabbatical year and the federal investigation, yet wasn't, was never place monetary gain ahead of patient care and benefit.

Finally, no one at the UU Medical School, to the best of my knowledge, went to jail for what they did because:

> *All that is necessary for evil to succeed is that good men do nothing.*
> Edmund Burke (1729 - 1797)
>
> and
>
> *The medical community has gotten more immunity than is healthy for society.*
> Rudolph Giuliani (b.1944)

****

The Saga of Dr. Goniff

At the same time I was battling, or should I say being victimized by, Drs. Kurveh and Mamzer, another battle was taking place between me and Dr. Goniff, a faculty member in another department in the Medical School. Mention is made of Dr. Goniff in the 4 December 1989 Report of the UU Retention, Promotion, and Tenure (RPT) Standards and Appeals Committee to the President of the University (Appendix C, Document 2):

> *2) A [negative] letter from [Dr. Goniff] was not provided to the Department of Pediatrics Promotion, Retention and Tenure (PRT) Committee, but was added to the file subsequently. This violates Faculty Regulations which specifically requires that the materials shall be submitted to the Department PRT Committee and makes no provisions for the addition of materials subsequent to their review other than recommendations from other appropriate parties and the candidate's response to these recommendations.*

Dr. Goniff's letter to the Department of Pediatrics Promotion, Retention, and Tenure (PRT) Committee, dated **31 October 1988**, stated in part:

> *Although I first met Dr. Fineman ten years ago...I did not have close contact with him until August of 1985, when I began visiting Utah on a regular basis. This contact, which increased once I took up a position here in May of 1986, revolved around our common interests in establishing a comprehensive birth defects registry, at both the state and regional levels. Accordingly, we have both served on committees...to investigate the feasibility of setting up a state-wide registry and, more latterly, to establish the protocols under which such a registry might operate. Our interaction also provided the catalyst for the two grants on which I am the Principal Investigator*

that will allow the development of a computerized birth defects/genetic disease data base in Utah (Eccles award) and for the region (Mountain States Regional Genetic Services Network award)....

Unfortunately, this initial productive interaction has not been maintained as I have found it increasingly difficult to work with Dr. Fineman. At this time a distinct rift has developed and I think it extremely unlikely that we will collaborate in the future. Despite this rift, I believe the following evaluation to be objective and fair. Nevertheless, the committee should be aware that Dr. Fineman and I no longer see eye to eye....

[In regard to his research,] *I feel Dr. Fineman cannot be characterized as a "leading investigator" in the field and does not meet the minimal requirements for attaining the rank of Full Professor. In fact, by comparison with the junior faculty in our department, his record of research would be rather marginal with respect to promotion from Assistant Professor to Associate Professor.*

[In regard to his teaching,] *while I have had little opportunity to observe Dr. Fineman in a classroom setting, I note that he is not responsible for developing a major course and that his teaching in confined to a smattering of lectures. Hence, I would judge that teaching does not represent a major contribution.*

[In regard to his clinical activities,] *although this is presumably an important area of Dr. Fineman's contribution to the University, I am not qualified to comment on his bedside clinical expertise. However, as far as I know he does not have a national reputation for his clinical skills.*

> [In regard to his administrative activities,] *I find this category to be the most difficult to evaluate, largely because of what is stated in his C.V. It appears from his C.V. that he is responsible for generating a large portion of the funds required to support the Division of Medical Genetics. Since [Dr. Kurveh], rather than Dr. Fineman, is responsible for the administration of the Division, I do not understand Dr. Fineman's role in managing and administering these major contracts.*
>
> [In summary,] *from my point of view - which is heavily biased towards scientific research - I feel Dr. Fineman's achievements fall considerably short of the minimal requirements for promotion to Full Professor. Since his scientific productivity seems to be diminishing, rather than the reverse, I do not see him attaining these requirements in the near future - if ever.*

At the same time I was seeking promotion, there was another faculty member who asked to be promoted to Full Professor in the Department of Pediatrics. He had one publication that did not describe a major scientific breakthrough, no grants of any kind, and he had not developed a major education course. He was granted his request for promotion, without difficulty. I am sure if Dr. Goniff were asked by the right person to write a glowing letter of reference for this other person he would have done so without hesitation - because a popularity contest by any other name is still a popularity contest.

It's also important to note that in his letter Dr. Goniff said he was the PI of the two grants we wrote. The reason he wrote this was because he was a thief and a liar.

6 November 1985: Part of a letter I sent to my 15-20 medical geneticist colleagues associated with the Mountain States Regional Genetic Services Network (MSRGSN)

> *Dear Colleagues,*
> *Would you please be so kind as to send me all the* [patient data collection/methods] *information you have from your state? I have been asked to attend a meeting in Washington, D.C., to discuss "specific facts regarding current reporting procedures for each state in your region, including similarities, differences, and problems."*
> ...

The MSRGSN (Arizona, Colorado, Montana, New Mexico, Utah, and Wyoming) was one of ten regional genetics networks in the U.S. funded by our federal government. The goals of the MSRGSN included, in part, improving the quantity, quality and accessibility to genetic services in the region; educating healthcare and social services providers, patients and families, and the lay public about medical genetic services; and improving patient data collection and evaluation methods. Before Dr. Goniff began his stay at UU, I had volunteered to be in charge of the regional data collection effort, and I was also a founding member and the co-director of the Network. **DR. AH** from the Colorado State Department of Health was the Director of the MSRGSN.

26 May 1987: Part of the cover letter of a $150,900, two-year grant proposal entitled, *State of Utah Comprehensive Program of Pregnancy Outcome Grant*, submitted by Dr. Goniff and me to the Willard Eccles Charitable Foundation of Salt Lake City

> [Introductory paragraph followed by] *Currently there is no centralized birth defects registry in Utah. We are requesting the Willard Eccles Charitable Foundation to provide start-up monies necessary to establish such a registry. The registry will be the first phase of a comprehensive follow-up care program that will benefit children, families, healthcare providers, and scientists in the years to come.*

> The establishment of this registry will offer assistance to families affected by hundreds of different birth defects and developmental disabilities and could eventually lead the way to the discovery of new ways of treating, counseling and preventing birth defects.
>
> The $150,900 grant we are requesting would pay for the first two years of the program. The Utah Department of Health has agreed to continue financial support once the program is on line. The money we are seeking will purchase the computer equipment and personnel necessary to start this comprehensive project....

Although Dr. Goniff's primary appointment was in another department in the Medical School, he spent a considerable amount of time performing research with members of our Division of Medical Genetics. He began working at the UU in May of 1986. He seemed like a decent person, and I never considered collecting reference information about him from the people who he had worked with previously. That was a major mistake on my part!

22 October 1987: A letter from the Willard Eccles Charitable Foundation to Dr. Goniff and me, with copies to several others at the UU Medical School, including the **Director of the Development Office**

> We are please to advise you that the Advisory Committee of the Foundation has approved your request of May 26, 1987, for $150,900 for the development of a core data base for the Utah Registry of Birth Defects and Genetic Diseases....
>
> As soon as you are in a position to proceed with the project, the Foundation will fund the budget for the first year described in your proposal.

I had worked for almost eight years to "line up the ducks" at the UU Medical School, including Dr. Goniff, with a huge amount of assistance from the leadership of the UDOH, the local chapter of the March of Dimes, and the Medical School Development Office to obtain the money and other resources needed to create and maintain a first-class Utah statewide registry of birth defects and genetic diseases. From my perspective, the single most important goal of the registry was to prevent human suffering, and premature death and disability, on a grand scale.

16 November 1987: A letter I received from Dr. AH, the Director of the Mountain States Regional Genetic Services Network

> *I am pleased to announce that the Mountain States Regional Genetic Services Network grant to the Colorado Department of Health has been approved ... for fiscal years 1988-1992. A* [Mountain States] *regional patient data collection system subcontract from the Colorado Department of Health to the University of Utah has been funded at the level requested, $60,768, for the first year.*
>
> *Please accept this letter as authority to begin the process to create the medical genetics patient data collection system....*

I had spent almost 900 hours over the previous two years advancing the cause of the MSRGSN at the Utah state, regional, and national levels. Almost all of this work was done without pay during weekends and holidays.

29 February 1988: A letter I received from Dr. AH

> *I am very happy to let you know that the subcontract between the Mountain States Regional Genetic Services Network and the University of Utah has been approved and signed to implement the Mountain States Regional Genetic Services Network computerized patient data base. I believe that I can speak for our entire network when I say how pleased I am that this subcontract has been completed.* ***In addition, I am delighted you are the principal***

> *investigator on this project. The time and effort over the last several years you have given to the development of this data base is very much appreciated and we will soon see the fruits of your work here in our region* [emphasis mine].

This letter documented in no uncertain terms who was to be the PI for the Mountain States Regional data base project.

Because Dr. AH played an important part regarding my dispute with Dr. Goniff, here's what Dr. AH wrote in his letter to Chairman Mamzer, dated **22 September 1988**, for my promotion to Full Professor of Pediatrics:

> *....First of all let me say that my acquaintance with Bob, although very positive, is limited to our close working relationship on the Mountain States Regional Genetics Network (MSRGSN) and the national Council of Regional Networks (CORN) for Genetic Services. Bob is a co-representative with me from the MSRGSN to CORN. He has acted in that capacity since the inception of MSRGSN in 1986. He has been and still is Chair of the Mountain States Network Data Committee which is one of the Network's most important, most active, and most productive committees. In addition, he is the representative from the MSRGSN to the CORN Data and Evaluation Committee. He is responsible for the subcontract between the Colorado Department of Health and the University of Utah for the establishment of the data base for our six-state network.*
>
> *Bob and I have worked rather closely in all the above activities and I have had a chance to observe him at many MSRGSN meetings (both committee meetings and entire network membership meetings) and numerous CORN meetings. In addition, Bob serves as a member of the Steering/Planning Committee of the*

MSRGSN and we have worked together at many meetings of this committee.

As I mentioned above, all of my contacts with Bob have been very positive. I am very favorably impressed with his administrative skills and his expertise in the field of medical genetics. In addition, to the best of my knowledge he is well regarded by other members of our Mountain States Network and CORN. As far as his independence and creativity are concerned, I have found that with a relatively small amount of guidance and direction from me and other members of our Steering/Planning Committee of the Western States Network, Bob is able to proceed on his own and to turn out an excellent end product on schedule.

In summary, my evaluation of Bob is very positive and is made without any reservations whatsoever.

****

By May, 1986, that is, during the first year of Dr. Goniff's tenure as a faculty member at UU, we had written several grant and contract proposals including one to the Willard Eccles Charitable Foundation in which Dr. Goniff was to be the PI and one to the MSRGSN for which I was to be the PI. There was no doubt whatsoever who would be the PI in each proposal.

Shortly after I received notification that the Mountain States data base proposal was approved, in the letter dated 16 November 1987, above, I contacted Dr. AH in Denver every few weeks to ask him when I could begin to spend money in the contract for the purchase of the necessary computer hardware, software, and data collection tools. Toward the end of January 1988, I was told by Dr. AH that the contract was in force and that Dr. Goniff was spending some of its money. I went to the office of **Robert (Bob) Glass**, the grants officer and an assistant dean of the UU Medical School to find out how, as PI, monies from the MSRGSN contract were being spent without my permission.

I had worked with Bob Glass on more than a half dozen grants and contracts during the previous decade. He was a bright, honest, knowledgeable, decent man who knew right from wrong. Bob told me Dr. Goniff made himself the PI of the MSRGSN contract - and then he showed me how Dr. Goniff had stolen the contract from me.

Dr. Goniff had assisted me in the writing of the MSGRSN proposal by providing, in total, 4-5 hours of discussion about it, most of which was during a dinner I treated him to at a local restaurant. After I had written the contract proposal, including its face or cover sheet known as the UU Document Summary Form, I showed the application to Dr. Goniff. This was a few days before the application was to be submitted to Bob Glass for the Dean's approval. Then the proposal would be sent to the Colorado Department of Health.

When I showed the contract proposal to Dr. Goniff, he said he wanted to read it one more time that evening, and he also offered to hand carry it to Bob Glass' office the next day. I said fine - no problem. In fact, what Dr. Goniff did that evening was forge a new Document Summary Form placing his name in the PI space. Bob Glass gave me a copy of Dr. Goniff's bogus Document Summary Form.

I immediately went to Dr. Goniff's office where I confronted him. I showed him a copy of his bogus Document Summary Form, and I insisted he transfer the PI position to me immediately. He said he "probably made a mistake" when he named himself PI on the Document Summary Form. He also refused to comply with my request. When I asked him why, he said he made himself PI because working on the contract would require more of his time than he originally thought, that is, an increase from two hours a week to four hours a week. Outwardly I did not lose control, but inwardly I was very angry! Only my upbringing saved Dr. Goniff from being thrown out the fourth floor window of his office that day. I was that angry and upset.

The next day I called Dr. AH in Denver and I asked him to send me a letter stating that I was the PI of the MSRGSN contract. I did not tell him why I needed the letter since my primary desire was to resolve the conflict with Dr. Goniff in a discreet and harmonious manner. At the time, I thought showing Dr. Goniff a letter from Dr. AH would speedily resolve our conflict. Unfortunately I was wrong. I showed Dr. AH's letter of 29 February 1988 (noted above) to Dr. Goniff, and I asked him to go with me to Bob Glass' office to correct the situation. He said no, again.

From that time on, Dr. Goniff no longer wanted to have much to do with me. At the end of March I met with him in his office to talk about the writing of the contract's progress report and renewal application. I was still the Chair of the MSRGSN data base committee, and I had a right to know what was happening/going to happen with the contract. Dr. Goniff told me he was too busy to talk to me, and I should come back some other time.

As mentioned above, during my conversation with Chairman Mamzer on 4 April 1988, I told him Dr. Goniff had stolen my MSRGSN data base contract. Chairman Mamzer responded exactly the way I expected. He told me rather abruptly I was paranoid, and I should see a psychiatrist.

In late April 1988, I talked several more times to Bob Glass and Dr. AH about ways to resolve my conflict with Dr. Goniff. Bob came up with what seemed to be a good idea. The MSRGSN contract had to be renewed before November, 1988. At the appropriate time, the Colorado Department of Health, under Dr. AH's direction, would send directly to me the official forms needed to renew the contract. I spoke by phone to Dr. AH, and he agreed to do this. At the annual meeting of the MSRGSN held in August 1988, I met privately with Dr. AH and he assured me the contract renewal information would soon be sent to me.

During the MSRGSN annual meeting, the various committees of the Network held their individual meetings. Dr. Goniff disseminated at the Data Committee meeting, of which I was still chairman, information about the UU contract and its renewal. Not only had Dr. Goniff removed me from the descriptive part of the contract renewal, he also removed the part that pertained to my salary and fringe benefits, and substituted instead 10% of his salary and fringe benefits. I was very concerned about this, and a few days after the annual meeting I called Dr. AH. He said I had no reason to be concerned.

It took a few weeks for Bob Glass and me to figure out that Dr. AH, Dr. Goniff and his chairman, the Dean of the Medical School, and several others at UU were not going to correct the situation because there continued to be honor among thieves, and because:

> *The world is a dangerous place. Not because of the people who are evil; but because of the people who don't do anything about it.*
> Albert Einstein (1879 - 1955)

22 November 1988: A letter I received from Dr. AH

> *We appreciate the work you did for the Network in your position as chairperson of the Data Committee. The most important thing we wish to do at the present time is to make sure the implementation of the Data System of the Mountain States Regional Genetic Services Network continues without interruption or delay. It is unfortunate that disagreements and misunderstandings have developed between you and [Dr. Goniff] and it is our hope these disagreements could be resolved. However, the subcontract for the Communal Data System and its related activities is with the University of Utah, not a specific individual. Since [Dr. Goniff] was the principal investigator of the contract at its inception, we see no reason to change his status at this time. It is the intent of this Network to fulfill its obligations as stated in the Mountain States Network grant and the contract with the University of Utah. Therefore, we would like to see the objectives and responsibilities of this contract implemented with as little interruption as possible.*

In other words:

> *It's not personal,* [Bob]. *It's strictly business.*
> Michael Corleone, *The Godfather* (1972)

****

The Saga of the Deans and the VP for Health Sciences

As the hierarchy was becoming more and more involved with my Medical School "issues," my new and more important emphasis became the hierarchy itself. What kind of people were the Dean and VP for Health Sciences at UU, and would they come to my rescue? I didn't have to wait very long to get the answers to these questions.

On 2 December 1988, I met with the Dean in his office to talk about Dr. Goniff. During the course of the conversation the Dean said three times Dr. Goniff "stole" my MSRGSN contract. He also said," Bob, we have to be pragmatic. Life is a bitch and then you die, and then they give away your clothing to Goodwill Industries." In response, what I should have said to the Dean, but didn't, was the Jewish adage, "Those who observe a thief and do nothing are also considered to be thieves." Remembering what previous UU Medical School Dean John Dixon allegedly said about med school deans, I thought to myself why waste time telling this guy he, too, is a thief? What I did say to him was Dr. Goniff was going to cause many more problems at the UU. Be prepared to deal with them.

Never get angry. Never make a threat. Reason with people.
Don Vito Corleone, *The Godfather* (1972)

19 January 1989: A letter the Dean of the Medical School sent to Dr. Goniff

> *I talked with [Dr. AH], Director of the Mountain States Regional Genetics Services Network, on the telephone this morning and he informed me that you had called him about the difficulty you foresaw in working with Bob Fineman. I am disappointed by your attitude toward working with Bob. I believe the work could have been accomplished if you had been willing to cooperate. I am disturbed because you did not talk to me before discussing this issue with [Dr. AH].*

> I believe that you usurped ["stole"] the contract from Bob when you signed the Document Summary Form making yourself the Principal Investigator and later when you refused to relinquish that position at his request. This act is inconsistent with the University of Utah Code of Faculty Responsibility and my expectations for School of Medicine Faculty members. I will place this letter in your file as part of the permanent record.
>
> [Dr. AH] asked me to give you the contract and I will comply with that direct request.

One of my friends gave me a copy of this letter. Otherwise, I never would have seen it.

Throughout my 13+ years at UU, this was the only time a person or committee mentioned the *Code of Faculty Responsibility*. I think that was because so few faculty members and administrators knew it existed. In addition, no one at the UU Medical School ever mentioned the oaths and/or codes noted in Appendix A. This most likely was because they thought the oaths and codes were not worth the paper they were printed on, unless of course it was toilet paper.

One of the criteria for promotion from Associate Professor to Full Professor was to be the director of a multi-centered, regional project. The contract Dr. Goniff stole from me fulfilled this criterion, and Dr. Goniff was only an Associate Professor. In July 1991, he was promoted to Full Professor. Was I surprised? Absolutely not, because, like I said before, ethical leadership, ethical decision-making, and ethical behavior were almost non-existent among the leadership of the UU Medical School.

The next several letters describe in greater detail how inept, incompetent, dishonest, and incorrigible the hierarchy of the Medical School was.

5 September 1989: A letter I wrote to the Willard Eccles Charitable Foundation, Salt Lake City

> On October 22, 1987, [Dr. Goniff] and I received a grant from the Willard Eccles Charitable Foundation to develop a core data base for the statewide Birth Defects Registry. With the Eccles grant now ending, I regret to report to you that [Dr. Goniff], the Principal Investigator, has not participated in the project for the last 8 - 9 months. The task of developing the core data base for the Birth Defects Registry has been completed by me and the coordinator of research and information systems, Utah State Department of Health….
> cc: Director of the School of Medicine Development Office

A few days after I wrote this letter, I told the Dean that Dr. Goniff hadn't done any work on the Eccles grant for more than eight months. I presume the Dean didn't know about my September 5th letter, above, to the Eccles Foundation because he responded, "Please don't mention this to anybody, especially the people from the Eccles Foundation."

11 January 1990: A letter from the Director of the Medical School Development Office to the VP for Health Sciences and the Dean of the Medical School

> I am forwarding for your review documents and correspondence related to the grant of $150,900 to Dr. Robert Fineman and [Dr. Goniff] from the Willard Eccles Charitable Foundation for the creation of the Utah Registry of Birth Defects and Genetic Diseases.
>
> Drs. Fineman and [Goniff] appear to have discontinued their collaboration early in the first year of this two-year grant. The information at my disposal raises reasonable doubt as to whether all grant funds expended by [Dr. Goniff] have been directed toward the creation of the Registry.

> *I have spoken with the coordinator of research and information systems at the State Department of Health, and am satisfied that the Birth Defects Registry now resides on their system, as indicated by Dr. Fineman in his final progress report of September 5, 1989 to the Eccles Foundation* [The Birth Defects Registry was supposed to reside in Dr. Goniff's UU computer purchased by the Eccles Foundation funds; except Dr. Goniff never created the computerized system. The only things he created were manure and smelly hot air].
>
> *The attached correspondence of our office from August through December, 1989, indicates that we have had difficulty obtaining a final progress report from [Dr. Goniff]. I am enclosing his draft final report, which I have asked him to prepare for submission.*
>
> *I am not competent to make a judgment on the progress reports submitted by Drs. Fineman and [Goniff], or whether they are irreconcilably inconsistent. Because I have reasonable doubt about the appropriate use of grant funds by [Dr. Goniff], I am passing these materials on for your review.*

The VP for Health Sciences and the Dean didn't respond to this letter according to conversations I had with the Director of the Development Office. In typical UU Medical School fashion, the matter was swept under the rug.

The Birth Defects Registry, on the other hand, was successfully implemented in the UDOH with only a small amount of input from Dr. Goniff. The $150,900 Eccles Foundation grant was barely used for its intended purpose. Of course, the money was misappropriated, or should I say stolen, by Dr. Goniff.

13 April 1990: A letter from the **Director of the Council of Regional Networks (CORN) for Genetic Services Data and Evaluation Committee** to me

> *Prior to January 1989, you were the official representative of the Mountain States Regional Genetic Services Network (MSRGSN) to the national Council of Regional Networks (CORN) Data and Evaluation Committee. I was favorably impressed by your plans to purchase computers and develop software, and collect data using a uniform data collection form. This plan was supported by MSRGSN [federal] funding.*
>
> *I am enclosing the results of [Dr. Goniff's] effort, as described in detail in the 1988 CORN data summary. In view of the very promising beginning and the generous support provided, the final results were very disappointing.*
> *cc: [Dr. AH] and the Dean of the UU School of Medicine*

While Dr. AH and the Dean never responded to this letter, I did.

26 April 1990: A letter I wrote to the Director of the Council of Regional Networks (CORN) for Genetic Services Data and Evaluation Committee

> *I read your letter of April 13, 1990 with great interest. I share your concern about the "disappointing" amount of data submitted to CORN last year by the Mountain States Region. My report for calendar year 1987 noted 97,009 laboratory tests and 11,006 patients. The report [Dr. Goniff] produced for you for calendar year 1988 noted only 44,006 laboratory tests and 1,038 patients. This clearly represents gross under-reporting and limits the value of the data.*
> *cc: [Dr. AH] and the Dean of the UU School of Medicine*

Once again, the matter was swept under the rug. Neither Dr. AH from the Colorado Department of Health nor the Dean responded to my letter.

About 18 months after this letter was written (and I had moved to Seattle), Dr. Goniff was replaced as the head of the MSRGSN data collection committee. The data collection effort I began was never finished because Dr. Goniff stole the contract. Hundreds of thousands of taxpayer dollars were improperly spent by Dr. Goniff. The hierarchy of the Medical School had many opportunities to stop and/or correct Dr. Goniff's unethical and illegal actions, but it chose not to do so.

What you have read so far is only a glimpse of what had happened at the UU. According to many of my physician friends, similar events and issues were occurring at same time at other medical schools and in private practices in the U.S. If you think these kinds of situations/events do not directly affect you, Google: "doctors"+ "self-interest," and/or: "doctors"+ "greed."

The Dean, Vice President for Health Sciences, and other high-level bureaucrats at UU were experts at sweeping excrement under the rug to avoid dealing with it. Other tactics they used frequently to avoid solving problems involving Drs. Kurveh, Mamzer, and Goniff were passing-the-buck, lying, and being duplicitous.

22 April 1988: A letter I received from the Dean of the Medical School

> *As I mentioned to you in our meeting this morning, the specific University committee to which you take your grievances is the Academic Freedom and Tenure Committee (AFTC)....As a tenured faculty member of this institution, you clearly have the right to petition the AFTC if you do not feel you have been fairly treated and you have not been able to adjudicate your differences with the department chairman and division chief after you feel you have made all reasonable efforts to do so.*

The Dean should have had a sign on his desk that said: **THE BUCK STARTS HERE.**

10 May 1988: A letter I received from the Dean of the Medical School

> *I am responding to your verbal request of last week concerning a meeting with you, [Chairman Mamzer], [Dr. Kurveh], and me. [Chairman Mamzer] informs me that he does not believe that a meeting would be fruitful at this time. You might want to discuss this matter further with your division chief, [Dr. Kurveh]. I, frankly, do not see any purpose in mandating a meeting and feel it is likely to be unproductive unless all parties involved are interested in further dialogue. As I mentioned to you in my previous correspondence, I do believe that you have further avenues to pursue if you do not believe your concerns have been properly addressed by either the department chairman or the dean.*

Even though I tried hard to resolve our differences at the Medical School level, it was not meant to be. After almost two additional years of strife the Academic Freedom and Tenure Committee (AFTC) issued its final report - see Appendix C, Document 3, the **12 March 1990 Report from the Chair of the AFTC to the President of the University of Utah**.

During my conversations with the Dean, he said he was distressed that Chairman Mamzer and I, two valuable members of the Department of Pediatrics, were not getting along, and that he was not going to "take sides in all of this." He told me this several times. At first I believed him, but not for long because he was incapable of telling the truth.

17 May 1988: A letter from the Dean of the Medical School, who was now also the VP for Health Sciences, to Dr. Kurveh

> *Thanks very much for meeting with me. I appreciate your approach to Bob Fineman and commend you for your even-handedness and*

> *integrity. While I do not look forward to the time consumed and the complications created by Bob's appeals, I appreciate very much your point of view that this may clear the air. At any rate, I am grateful to you for all that you do and stand ready to help you and [Chairman Mamzer] in any way I can.*

The Dean/VP for Health Sciences unquestionably lied to me. If he was really interested in saving time and energy he should have sacked Chairman Mamzer. Think of the enormous amount of ignorance, grandiosity, and/or arrogance it took to write a letter like this, because I'm sure he never thought I would see it.

As I mentioned before, I had many concerned and supportive friends at UU; for example, in the Department of Pediatrics including in the Chairman's office and in other departments at the Medical School including the Dean's office. I also had friends throughout the main campus of the University, including current and past Vice Presidents. They supported my cause and, to this day, I am very grateful for what they did. I knew I was in a war, so-to-speak, and one of the things I needed most was good information. Fortunately, there were many people who were happy to give me copies of letters like the ones above.

1 July 1988: A letter I received from the Dean/VP for Health Sciences

> *I received your note yesterday concerning your decision to present your case to the Academic Freedom and Tenure Committee. I support your decision, albeit with misgivings that I have expressed to you previously.* ***I appreciate your sense that I have been even-handed in the matter. That is my complete intention and will continue to be*** *[emphasis mine]. I am distressed that good friends and colleagues are not able to resolve their differences amicably and will of necessity spend time and energy in this way rather than on productive activities.*

I never in any way indicated to anybody that the Dean/VP for Health Sciences was acting even-handedly. He didn't know I had a copy of his 17 May 1988 letter to Dr. Kurveh, and I had no reason to tell him about it. First, if I told him, nothing about him would have changed and, second, I would have gotten one of my friends in trouble. The Dean/VP, whose actions regarding me were both immoral and amoral, treated me like a mushroom. He kept me in the dark about his true intentions and actions, and he fed me a lot of manure.

****

As an aside, there are several other reminiscences I need to mention at this time that could explain the causes of some of doctorhood's major deficiencies.

In 1987, faculty members from different departments in the UU Medical School and elsewhere at the University, Michael and Phyllis Walton, and I created an elective course for 4th year medical students. The 10-week course emphasized non-scientific aspects of medicine such as medical ethics, law, history, anthropology, and policy. As faculty members, or in the case of the Waltons, interested parties, we thought medical students should know more about medicine than what was traditionally taught in most medical schools. Another faculty member, Dr. Charles Hughes, a medical anthropologist, and I were the course co-directors.

Of the 100 seniors in each of the two 4th year classes that could have enrolled in our elective course, that is, during the 1988 - 1989 and 1989 - 1990 academic years, only about ten in each senior class signed up for the course. I was puzzled as to why this was so. I asked about a dozen students from each senior class in a non-scientific way why so few of them had signed up. Almost uniformly I got the same response, "If your elective isn't going to help me make money or pass the medical boards [the national, standardized tests which have to be passed to obtain a medical license in every state in the U.S.], I'm not interested."

In fact, the students were right. The course was not designed to help them make money or pass the medical boards. It was being taught to help them understand there was more to becoming a doctor than making money, and more to a medical education than passing the Boards. Alas, my words of wisdom, at least I thought they were words of wisdom, fell on deaf ears as I subsequently tried, without success, to interest more students in taking the course. In addition, their comments reminded me of the time when I asked my freshman med school classmates, in 1966, why they wanted to be doctors: the most common answer - to make money, of course.

Around the time our senior elective course was being taught, the Dean/VP for Health Sciences asked me to meet him in his office. He showed me a letter he had recently received from a physician in private practice in Salt Lake City. The physician wrote that practicing medicine in Salt Lake was like working in a "jungle" because doctors were "stealing" each other's patients, and arguing incessantly over money. The physician asked the VP what they could do together to help improve the situation. After we talked for about 10-15 minutes, the Dean/VP concluded the conversation by saying there was nothing he could do to help the situation. I've always been mystified by the purpose of the meeting because the letter was written by a local physician to the Dean/VP. It was never explained to me why the Dean/VP took the time to discuss the letter with me.

Several years later, when I was working for the Washington State Department of Health and living in Seattle, I talked to a man who is considered to be one of the grandfathers of the medical school-medical ethics education movement in the U.S. He told me about his attempts to implement a medical ethics course in the late 1960's and early 1970's at the school where he worked.

First, he went to the Dean and explained what he wanted to do. The Dean thought teaching medical ethics to the students was a great idea. He said he supported the effort. A presentation was made at one of the regularly held <u>basic sciences</u> chairmen's meetings. All of the basic sciences chairmen thought it was a great idea until they were asked to give up a total of 25 hours of teaching time, during the <u>first two years</u> of medical school. Not one chairman was willing to give up even one hour of their allotted teaching time.

A presentation was then made at one of the regularly held meetings of the chairmen of the clinical departments. All of the chairmen thought it was a great idea until they were asked to give up a total 25 hours of teaching time, during the second two years of medical school. Not one chairman was willing to give up even one hour of their allotted teaching time.

It took several more years, a lot of persistence, and a crowbar for my friend to implement a 25-hour course in medical ethics at the school.

****

The President's Saga

When you read the following articles about the cold fusion debacle at the University of Utah, remember the President of the University was the physician/circus master who ultimately held my fate in his hands.

28 MARCH 1989: "Cold Fusion Experiment May Lead to Big Bucks for the University" *The Daily Utah Chronicle*, the student-run newspaper of the University of Utah

> *In the aftermath of Thursday's startling announcement of cold fusion by a University of Utah professor and a colleague, U officials are now scrambling to control the legal side of a discovery that could possibly alter the future of energy worldwide and mean a huge shot in the arm for the state's economy....*

2 May 1989: "Failed Attempts to Repeat University of Utah Experiment Are the Only Results in the Utah Alchemy Circus" *The New York Times*

> *For the last month, scientists around the world have been poised between the deepest doubt and highest hope. The University of Utah claimed on March 23 that two researchers had learned how to produce nuclear fusion at room temperature. Yet despite a month of attempts to*

> repeat the experiment, no one knows yet if the claim will evaporate in smoke and recrimination or prove the first step to a revolutionary new source of energy....
>
> Last week, [the] President of the University of Utah appeared before a U.S. House of Representatives committee to drum up federal funds. Asked how much, he replied, "The figure that comes to mind is $25 million." Given the present state of evidence for cold fusion, the government would do better to put the money on a horse.
>
> For the two University of Utah scientists, the best bet is to disappear into their laboratory and devise a clearly defined, well-understood experiment that others can reproduce. Until they have that, they have nothing. As for the University of Utah, it may now claim credit to the artificial heart show and the cold fusion circus, two milestones at least in the history of entertainment, if not science.

The master of ceremonies of the "artificial heart show," which took place at the UU Medical Center in the early 1980s, was none other than the President of the University who, back then, was the VP for Health Sciences.

4 February 1990: "What the Experiment Produced" *Newsday Magazine*

> Even without calling for eye of newt and toe of frog, the recipe for cold fusion was strange enough to set eyes rolling and teeth grinding. But it was much too tempting not to try....
>
> As a result, it may be a long time before the University of Utah regains respect, especially among scientists who at first were dumbfounded by the news, then became skeptical and are now visibly hostile....

Scientists at Caltech called the UU cold fusion experiment the result of incompetence and delusion by its two authors. A physicist from CERN (the European Organization for Nuclear Research) called the entire episode an example of pathological science.

8 August 1989: "University Begins Moving Into Cold Fusion Research Institute" *The Salt Lake Tribune*

> *The University of Utah began moving equipment into its Cold Fusion Institute in Research Park Monday after the state's Science Advisory Council approved its $4.5 million, two-year budget. Most council members and university officials admitted they were taking a gamble, but they said the potential return of the experiment, which could lead to a new energy source, made it well worth the risk. In the end, the two scientists on the nine member panel were the only ones who did not go along....*

1 June 1990: "'External' Cold Fusion Funds Actually Came from Within University" *The Salt Lake Tribune*

> *A $500,000 "anonymous" donation to the University of Utah Cold Fusion Research Institute, the largest outside funding the state-supported institute has received, actually came from another entity within the university.*
>
> *The revelation prompted disdain from members of the panel that oversees the state's $5 million research investment. And, 22 members and the dean of the College of Science issued a joint statement calling the situation an "apparent deception" and asking that a complete financial audit and scientific review of the institute be carried out....*

2 June 1990: "[U President] says 'Palace Revolt' at the University Won't Threaten His Job" *The Deseret News*

> The main question being asked Friday on and off campus was why hadn't the source of the $500,000 gift from the University Research Foundation been named at the beginning? "That's the question I'm grappling with today," [the President] said. "The answer is that most of the time when money is allocated within the University we don't describe the sources, partly to avoid a level of controversy. Cold fusion is an area of high visibility and controversy, and I thought it would simply add to the controversy. That is the way it turned out. It has added triply to the controversy." But the controversy was compounded Friday by confusion over when it was known that the University itself was the "anonymous" donor.

The President of the University and his central administration didn't understand that they could deceive some of the people some of the time, and get away with it; but they couldn't deceive all the people all the time, and get away with it.

When you read the following information about the bombshell concerning the re-naming of the UU Medical Center, note that it went off only a few months after the cold fusion fiasco hit the fan.

28 July 1989: "Legislator Joins Protest Against New University Hospital Name" *The Salt Lake Tribune*

> A well-known donor and a state senator have joined the ranks of those protesting the naming of the University of Utah medical school and medical center after James Sorenson, a businessman who gave the institution $15 million.
>
> Joseph Rosenblatt, a university donor since the 1940s, said he helped draft a new petition protesting the name change, and he'll do

> *whatever else is necessary to retain the old "University of Utah Medical School" name....*
>
> *The newest petition protesting the name change calls for a building or an endowment fund to be named after Mr. Sorenson - not the medical school. [The President of the University] said he negotiated the contract with Mr. Sorenson to ensure excellence at the school and to honor all supporters, including Mr. Sorenson. Last spring the University's governing Institutional Council voted 7 to 3 in a closed door meeting to rename the medical school and medical center in exchange for the $15 million gift. The split vote proved to be a harbinger of divisions occurring now within the university and community on the name change.*

1 August 1989: "Critical Senate Finds Sorenson Debate an Untimely Issue" *The Salt Lake Tribune*

> *The State Senate placed the newly named James Sorenson School of Medicine on its critical list Monday by protesting removal of the school's "University of Utah" identity. Senators were even poised to pass a resolution denouncing the renaming, which university officials took upon themselves after a local entrepreneur recently pledged $15 million in stock to the school....*

3 August 1989: "Sorenson Lets Medical School Keep Money, Name" *The Salt Lake Tribune*

> *Businessman James Sorenson released the University of Utah Wednesday from a contract naming the medical school and medical center after him while allowing the university to keep his $15 million gift. In less than two weeks, more than 1,200 doctors, nurses, medical students and staff signed a petition protesting the name change....*

2 September 1989: "Return $15 Million Donation, Sorenson Asks University" *The Salt Lake Tribune*

> *Businessman James Sorenson rejected the University of Utah offer to name a research center after him instead of the originally proposed university medical school, and asked the university Friday to give back his $15 million gift. Sources told The Tribune that [the President] had considered resigning as university president* [I should have been so lucky] *if a compromise couldn't be reached with Mr. Sorenson, but after several discussions with supporters, [the President] decided to keep his post.*

It became obvious to me in the summer of 1989 I was working in a cesspool of incompetence and dishonesty. My primary objective then was to leave the UU intact: physically, emotionally, financially, spiritually, and otherwise; that is:

> *You got to know when to hold 'em, know when to fold 'em, know when to walk away and know when to run.*
> Kenny Rogers (b.1938)

Taking my complaints to the highest levels was a prerequisite for any future lawsuit I could have filed against the University, because no appellant had ever won a case similar to mine on appeal in a civil suit unless he/she had exhausted all available avenues at his or her university.

Bonnie and I had many wonderful friends in Salt Lake City and we really did not want to leave. We thought about staying there, however, I had burned too many bridges, and the Medical School had too many close ties to the local medical community for me to get another job there. Unlike New York, Philadelphia, Baltimore, or Boston, there was/is only one medical school in Utah.

5 June 1990: "[U President] Gets No-Confidence Vote" *The Deseret News*

> *After what amounted to a vote of "no-confidence" by the faculty, the fate of [the] University of Utah President was in the hands of the State Board of Regents Tuesday. And the Regents won't decide for more than two weeks whether they want to evaluate the University president's leadership. Faculty members are increasingly unhappy with [the President] for his handling of cold fusion, his decision to rename the University Medical Center after James Sorenson, and the elimination of the Office of Academic Vice President.*

6 June 1990: "Donor Urges [U President] to Open Up" *The Salt Lake Tribune*

> *[The U President] shouldn't resign as President of the University of Utah but he must reverse his closed-door policy and listen to faculty members and college deans, a major University donor said Tuesday. Joseph Rosenblatt, who is also a close friend of [the President], said the president's biggest mistake may be "his inability to work with people who are his equals...." Mr. Rosenblatt said the university president mistakenly promoted good teachers to administrative posts, "who have no more ability to manage than a man who runs a hotdog stand. [The President] has a vision but he must get good people who will give him good advice."*

Joe Rosenblatt was a very knowledgeable, long-time supporter of the University, financially and otherwise. However, in this instance, he was only partly right. People who run hotdog stands were/are, in many ways, superior to many of the individuals who served in the University's central administration. Also, the Waltons and I tried several times to give the President good advice about the hierarchy at the Medical School, including Chairman Mamzer, the Dean, and the VP for Health Sciences.

The President refused to listen to us because, in part, he couldn't figure out who his real friends were.

Knowing my time at UU would soon be over, it was necessary for me to obtain the support of a very powerful, outside ally to finally defeat the hierarchies of the Medical School and University. Enter the fray the Washington, D.C.-based, American Association of University Professors.

19 December 1989: A letter from the **Associate General Secretary of the American Association of University Professors (AAUP)** to the President of the UU

> *Dr. Robert Fineman has for some months been in consultation with this Association about his rights under generally accepted principles of academic freedom, tenure, and due process. He has recently shared with us the December 4 report to you from the University RPT Standards and Appeals Committee [Appendix C, Document 2], on his appeal against rejection of his candidacy for promotion to the rank of professor.*
>
> *As you know, that Committee, having examined the available documentation and having held a formal hearing, found an array of serious errors of procedure in the evaluation and concluded unanimously "that the review process was in fact deeply flawed and heavily biased against the appellant."...*

The RPT Standards and Appeals Committee was a very important and powerful standing committee. Its findings and recommendations were reported directly to the UU President, and it was the Committee of last resort/final appeal regarding retention, promotion, and tenure issues. There were no perks for being a member of the Committee - only a multi-year, voluntary commitment aimed at serving the best interests of the University.

Most of the members of the RPT Standards and Appeals Committee were affiliated with other programs at the University, not the Medical School. Committee members did not understand the culture and mentality in the Medical School and, especially, its administration. What most members of RPT Standards and Appeals Committee failed to understand was I had as much chance of receiving a fair promotion review in the Medical School as Tom Robinson had at trial in *To Kill a Mockingbird* (1960). This was because there was among many of the leaders in the Department of Pediatrics, the Medical School, and the University, overwhelming hubris and a lack of morality.

5 January 1990: A letter the VP for Health Sciences sent to me

> *I am writing in response to the request of [the President] that I conduct a completely new review of your request for promotion. Because I take seriously the instruction to "take exceptional precautions to ensure University procedures are followed and that both the interests of the University and your interests are protected," I am actively pursuing those matters....*

After I read this letter, I put on my hip waders because I knew the VP was doing his incompetent, manure-spreading routine again!

7 March 1990: A letter I sent to the VP for Health Sciences

> *Upon first reading your proposed promotion procedure dated January 31, 1990, I had the impression that you were unacquainted with the report which occasioned [the President's] request to you. That report detailed the ways in which the promotion process was subverted by [Dr. Mamzer], the Department of Pediatrics promotion committee's bias, and the inaction of the Dean of the School of Medicine. It seems to me that any further involvement of these individuals and that committee in my promotion review would run counter to the findings of the report. Indeed, UPTAC [the University Promotions and Tenure Advisory Committee] last year concluded that I could not get a fair hearing in the Department of Pediatrics.*

> *Given the findings of both UPTAC and the RPT Standards and Appeals Committee, the procedures you outlined in your letter of January 31, 1990, are inadequate to ensure a fair and unbiased promotion review. I would be happy to meet with you to discuss how the issues of fairness raised by the RPT Standards and Appeals Committee could be satisfactorily addressed in a review process.*

7 March 1990: A letter the VP for Health Sciences sent to me

> *Your letter of March 7, 1990 was received by me today. Your objections are recorded and will be considered. Your perceptions differ markedly from mine, and I have to admit great disappointment that it is your view that important steps in the review process for promotion can be summarily disregarded. I emphasize that this is only my perception and will, therefore, need to consider your position carefully and seek considerable consultation on the matter* [in other words: **THE BUCK STARTS HERE**].

7 April 1990: A letter the VP for Health Sciences sent to me

> *I believe that you have recently received a copy of [the Provost's] April 5 letter to me concerning proposed procedures for a second review of your application to be promoted to Professor of Pediatrics.*
>
> *It is my decision at this time to proceed with the review with the procedures as I have outlined them, although I am requesting that [Dr. Mamzer] not participate in the review and, instead, appoint an associate chairman to fill that role. The values and traditions of academic freedom are simply too important to summarily exclude the steps and groups that have important responsibilities and interests in the process.*

> *Because I believe that this should be a collegial process, it is my suggestion that we meet together without other representatives or advocates. Should you decide that such a meeting is not in your best interests and wish only to meet with your representatives present, then by definition the proceeding becomes adversarial, and I will not meet with you without legal counsel present as well....*

16 April 1990: A letter I sent to the VP for Health Sciences

> *It is clear from your correspondence with me that you have been unable to meet the concerns of the RPT Standards and Appeals Committee. The procedure you proposed is totally inadequate to deal with the following problems:*
>
> *[Dr. Mamzer], whose bias and malice have been manifest throughout the last several years is still involved in the promotion process;*
>
> *You have proposed no way to ameliorate the damage that [Dr. Mamzer] has done in contacting people solicited for review letters or the problem of the significant amount of time usurped from my professional pursuits to protect my interests within the University;*
>
> *[Dr. Mamzer], with the acquiescence and even the encouragement of the administration, long ago turned my promotion procedure into an adversarial proceeding. Yet I am now expected to pretend that this is not the case; and*
>
> *[Dr. Mamzer's] campaign to force me out of the University, including cutting my salary by 30% and removing me as director of the chromosome laboratory, has interfered with my professional career. In essence, two years of my professional life have been taken away from me.*

> *If you can think of no way to meet the concerns of the RPT Standards and Appeals Committee, and your proposal does not do so, I request that you either: 1) return the matter to the RPT Standards and Appeals Committee with a request for help in establishing a fair procedure, or 2) immediately recommend to the University administration my promotion to full professor.*
>
> *Regardless of the presence of my advisor or advisors, I suggest that [a lawyer] from the Attorney General's office be included in any future discussion between us. Perhaps [he] can more clearly define and protect the interests of the University, including its faculty - more than the administration has yet been able or willing to do.*

16 April 1990: A letter from the Dean to all Medical School faculty and staff

> *I have just been informed that [the VP for Health Sciences] has resigned and will be leaving the University to join Intermountain Health Care....*

One by one the rats began to abandon the ship - starting with the VP for Health Sciences. I would like to believe I had something to do with their departures for the sake of the UU medical students, faculty, staff, and the patients and their families. On the other hand, would you believe this VP later became the president of another university? *Oy Vey!*

26 April 1990: A letter from the Associate General Secretary of the American Association of University Professors to the VP for Health Sciences

> *Dr. Robert Fineman has shared with us correspondence that has ensued since we wrote on December 19 to [the University President] with respect to the issue of Dr. Fineman's candidacy for promotion in rank. The key current issue, we take it, is whether an entirely new evaluation can realistically be expected to meet essential*

standards of fairness if members of the Department of Pediatrics who participated in the flawed previous evaluation are now to participate in the new evaluation. [The President], in his December 19 letter calling for the new evaluation and asking you to "take exceptional precautions" in conducting it to see to its soundness, sought assurance and that the ten concerns specified by the RPT Standards and Appeals Committee as having tainted the original evaluation "do not resurface in any significant way." You indicated in your April 7 letter to Dr. Fineman that you were asking the chair of the department to step aside in the new evaluation, but that your current decision was to invite him to appoint someone to substitute for him and to include the department's promotion committee again in the evaluating process.

The RPT Standards and Appeals Committee's stated concerns implicate other members of the Department of Pediatrics in addition to the chair. In your letter to Dr. Fineman, you referred to the importance of "the values and traditions of academic freedom" as reason for again including the department members. We would suggest to you that the focus, in terms of upholding principles of academic freedom in Dr. Fineman's case, should be ensuring, pursuant to [the President's] charge, that there be no repetition of the assaults on due process by department members that occurred during the original evaluation. We ask you to consider bringing the matter of Dr. Fineman's candidacy for promotion to a decision without assigning any further active role in the process to members of the department who were previously involved.

There were times when I felt sorry for the two Medical School Deans I dealt with, the VP for Health Sciences, and the President, because their actions were inept and incompetent, and they were in it way over their heads - not only about the issues that pertained to me, but also in regard to the other problems described above. Even though I thought they were behaving unethically, immorally and/or incompetently, I still had very mixed feelings about them.

25 May 1990: A letter a Salt Lake City lawyer friend of mine wrote to the President

> *In our representation of Dr. Robert Fineman, we request a meeting with you and your legal counsel by June 10, 1990. Although the University RPT Standards and Appeals Committee issued its report on December 4, 1989, and the Academic Freedom and Tenure Committee made its recommendation on March 9, 1990, the University has taken no remedial action toward Dr. Fineman. The delays have had catastrophic effects on Dr. Fineman's career and economic security. They also breach his contract and tenure rights.*
>
> *Failure to respond to this request will operate as a formal declaration of the University's rejection of its own due process procedure and will leave Dr. Fineman no option but to seek a remedy through the courts.*

11 June 1990: "The Period of Time I Can Effectively Provide Leadership is Nearing an End" *The University of Utah Quarterly Review*

> *This operational phrase, delivered crisply and unemotionally at a press conference, was the denouement of a year in which the president, while not under siege, faced an increasingly fractionalized faculty....*
>
> *[The President] was like a person astride two railroad tracks that had inched apart. He either had to bring the rails back to parallel or get off. Within a week's time in early June, he was the object of the Academic Senate's overwhelming vote for a "review" of his ability to lead the*

> University; the focus of undiluted apprehension at the triumphant conclusion of a five-year fund-raising campaign; and the subject of media scrutiny and community surprise when he called a news conference to announce his decision containing the statement above. He added, "It is my intention to position this university for new leadership following the 1990-91 academic year. At that time I will retire."

Even in the environment that was the University of Utah, sooner-or-later the chickens had to come home to roost:

> *And yet we cannot call it valor to massacre one's fellow-citizens, to betray one's friends, and to be devoid of good faith, mercy, and religion; such means may enable a man to achieve empire, but not glory.*
> Niccolo Machiavelli (1469 - 1527)

On 16 July 1990, about five weeks after the President announced his resignation, the initial chunks of manure hit the fan regarding Pediatric Faculty Physicians Inc., as described above in **The Saga of Dr. Mamzer**. Obviously, it would have been better if Dr. Mamzer and the hierarchy of the University had been running a hotdog stand, because then it would have been only a hotdog that hit the fan!

21 August 1990: A letter from the Associate General Secretary of the American Association of University Professors to the UU President

> *We wrote you last December 19 with respect to the report and recommendations of the University RPT Standards and Appeals Committee in the case of Dr. Robert Fineman and his candidacy for promotion in rank. We wrote again on April 26 to [the Vice President for Health Sciences], pursuant to the RPT Standards and Appeals Committee's findings, that the matter be decided "without*

assigning any further active role in the process to members of the Department of Pediatrics who were previously involved."...

Our expressed concern over a further role for the previously involved department members stemmed not only from the RPT Standards and Appeal's Committee's report but also from the reported belief of the members of the University Promotion and Tenure Advisory Committee [UPTAC], upon noting the intrusion of personal issues into the departmental evaluation of Dr. Fineman's candidacy, that remanding the case to the department would not lead to a fair hearing. We regret very much that the central administration of the University of Utah has not come forth with a course of action that would serve to bring this matter to an appropriate resolution.

Our concern relating to Dr. Fineman's situation is compounded by the administration's apparent failure to act on the recommendations of the University Academic Freedom and Tenure Committee [AFTC]....

If the administration has compelling reasons for rejecting the recommendations of the Academic Freedom and Tenure Committee, it should set them forth and ask for reconsideration. The AFTC reports, as well as the University RPT Standards and Appeals Committee report, impress us as thoughtful and thorough, the result of careful examination of the fullest record obtainable by fair-minded academic people with a high sense of responsibility. For the administration to decline to be responsive to these two duly constituted University bodies seems to us to be inimical to basic principles of academic government and academic due process.

> As we consider pursuing our interest in these important matters under our longstanding responsibilities, we would welcome your comments.
>
> cc: Utah Commissioner of Higher Education
> University of Utah Institutional Council
> Vice President for Health Sciences
> Vice President for Academic Affairs
> Chairperson, Academic Freedom and Tenure Committee
> Chairperson, RPT Standards and Appeals Committee
> Dr. Robert Fineman

The Academic Freedom and Tenure Committee was another important and powerful, UU standing committee. Its report of 12 March 1990 (Appendix C, Document 3) confirmed what I had thought about the doctors I had been dealing with. They had no respect for the University and its rules.

As noted in *II. Private Practice Income (PPI)* of the AFTC's report, there were rules and regulations regarding PPI that Drs. Mamzer and Kurveh disregarded, mostly for punitive reasons. Disregarding rules and regulations, and not being held accountable for such actions, were attributes that were common at the Medical School. One of the reasons was: <u>We're doctors and we don't have to abide by rules and regulations because we're above those things!</u>

In regard to *III. Academic Standing*, it was mostly my fault I lost the "debate" regarding this point. I should have done a better job presenting information to the AFTC regarding my clinical/patient care and teaching responsibilities, in addition to overseeing the chromosome laboratory.

There was no effective system of checks and balances in the Medical School that protected faculty members when disagreements occurred regarding job descriptions or other issues like promotion and retention. <u>Past promises by Chairs, even when they were in writing, often meant nothing. Deception was commonplace. Administrators and faculty members would say</u>

<u>one thing one day, and say or do the exact opposite the next. Chairmen frequently acted as though they were omnipotent. They often acted as though they could do anything they wanted - even destroy the careers of others, without recourse.</u>

Most of the members of the AFTC were not faculty from the Medical School. They did not understand the differences between the Medical School and the rest of the University. For example, in the Medical School a person could hold the position of Chair for life. On the main campus, chairpersons, usually Full Professors, rotated in and out of the position every few years. Chairs on the main campus, therefore, were not nearly as powerful or controlling.

Despite the fact I had lost the *III. Academic Standing* vote, I was very happy with the AFTC's report, and I thanked its members for their time and effort.

Shortly after the AFTC report was rendered, I talked to its Chair and I asked her to tell me more about *III. Academic Standing*. I did not argue with her or question the Committee's decision. She said that if a Chairman "ran his department into the ground", or if the President "ran the University into the ground, there's nothing the AFTC could do about it." <u>She also told me the AFTC received more complaints from the Medical School than all the other schools and units at the University, combined!</u> It's no wonder university presidents say their biggest headaches come from "docs and jocks." I told her many innocent people had already suffered at the hands of Dr. Mamzer and, if he was left unchecked, many others would suffer in the future. I did not realize she would be one of them. Several months later, Dr. Mamzer became a member of the University Promotion and Tenure Advisory Committee (UPTAC), and tried to block the AFTC Chair's promotion to Full Professor in the English Department.

****

29 August 1990: A one-hour meeting took place that included the new Vice Presidents for Health Sciences and for Academic Affairs, the University's lawyers, the Waltons, and me.

I presented a list of 15 conditions necessary for me to continue my career at UU including, in part, my switching to another department in the med school and taking with me the chromosome laboratory and the birth defects registry computer purchased with the Eccles Foundation money; my private practice income (PPI) would be restored from the time it was cut to the present; my promotion to full professor; and Drs. Mamzer, Kurveh, Goniff, and CD be told to cease making defamatory remarks about me and be excluded from any future evaluations or decisions regarding me.

At the end of the meeting, it was obvious that the University's central administration and its lawyer had no intention of agreeing to my conditions. In fact, neither the Waltons nor I ever thought they would. Two additional options were discussed: an out-of-court settlement and a lawsuit. With regard to a lawsuit, the University lawyer said that if one took place, "There's going to be a lot of blood on the floor." Michael Walton responded, "But none of it will be Fineman's." The lawyer's jaw dropped, but he didn't respond. Michael then said to the lawyer, as we had said several times before to others in the University's hierarchy, "Get rid of Chairman Mamzer before it's too late." No response, again.

12 September 1990: A letter from the UU President to the Associate General Secretary of the American Association of University Professors

> *In response to your communication of August 21, 1990 regarding Dr. Robert Fineman, I am pleased to report that significant progress is being made towards a resolution of Dr. Fineman's grievances. Specifically, in the last several weeks the Vice President for Health Sciences and our University's legal counsel have been involved in substantive negotiations with Dr. Fineman and his advisors toward the goal of a resolution of this issue which is satisfactory to Dr. Fineman's interests and those of the University of Utah. It is quite likely a resolution will be forthcoming soon...*

Negotiations took place during the middle of September. The UU Institutional Council voted to sustain an out-of-court settlement shortly thereafter. There was nothing in the settlement that prohibited me from writing this book.

13 November 1990: A letter I sent to the new VP for Health Sciences

> *I herewith resign my professorship in the Department of Pediatrics effective December 31, 1990....*

Part of the out-of-court settlement was my promotion to Full Professor of Pediatrics and, while both sides agreed not to discuss in public the amount of money involved in the settlement, a University representative disclosed the amount to the media shortly after I left UU. What I received was almost five-years of my full salary (base plus private practice income) which amounted to $300,000 in calendar year 1990 dollars (equivalent to $465,000 in calendar 2008 dollars). Where the money came from was very interesting and is discussed below.

While some thought or may still think I won a Pyrrhic victory, a victory won at an excessive cost, I don't believe it. I will always remember the good things I accomplished at UU, our friends from Salt Lake City who we still keep in contact with, and the cash settlement. We could not have purchased a home in Seattle without the money. In addition, when the actions of Pediatric Faculty Physicians Inc. were later found to be overwhelmingly disreputable, I didn't have to plead the Eichmann defense: "I was just following orders," because I had refused to have anything to do with PFP Inc.

According to a 30 December 1989 article in the *Los Angeles Times* entitled, "The Rewards of Conviction," the playwright Vaclav Havel (b.1936), in 1979, then a prisoner of conscience in a Czechoslovak jail, wrote in a letter to his wife, Olga:

> *There are times when an artist [or a physician] must put his art aside in order to do something positive in life, something modest that may not earn him a place in history, but which is an*

> *expression of a moral imperative or simply a love for people.*

I was no Vaclav Havel, nor did I spend time in jail. Still, I can say what I did was, for the most part, for the same reasons Mr. Havel went to jail.

The UU lost a real "rainmaker" when I was forced to leave. All of the genetics grants and contracts I wrote over the decade of the 1980's, totaling millions of dollars, were funded. And, after a two-year hiatus, from end of 1988 to the end of 1990, my career flourished again like it did from 1977 - 1987.

15 November 1990: A letter I wrote to all the full-time faculty members of the Department of Pediatrics, excluding Chairman Mamzer

> *As I depart the University of Utah, my thoughts turn to the organization of our Medical School. I can think of no better primer concerning medical school politics than Machiavelli's, "The Prince." Written in the 16th century, it describes the vices and virtues of princes, both real and ideal. Because the attitudes of Renaissance princes closely parallel those found in many academic medical centers, I enclose a copy of Machiavelli's masterpiece for your edification and confirmation.*
>
> *With best wishes for your future success I remain faithfully yours,*
> *Bob Fineman*
> *Enclosure - "The Prince"*

As I was heading out the door of UU, a friend, who was a fellow faculty member in the Department of Pediatrics, said to me, "Don't worry, Bob, there's life after the University of Utah." I smiled, said "Thank you," and thought to my self: *L'Chaim, L'Chaim!* - To Life, To Life!

3 May 1991: A letter a Division Chief from the Department of Pediatrics (not Dr. Kurveh) sent to me five months after I left UU

> *You may think that you have somehow struck back at [Dr. Mamzer] or the University of Utah. You have not. The settlement you received will be paid for by me, and by all of my colleagues out of clinical income and other precious, limited resources. My own division will be assessed tens of thousands of dollars to pay your settlement.*
>
> *I know very little about your case, and I know that you are personally wounded by what happened. That makes me personally sad because I think you are a fine individual and that the differences developed between you and your division, and eventually the Department of Pediatrics, were extremely unfortunate.*

It saddened me to learn that all 84 full-time members of the Department of Pediatrics were taxed in excess of $3,000 each by Chairman Mamzer to pay for my out-of-court settlement. One reason for this was the State's Risk Management Pool, the state-run insurance carrier, refused to pay for the settlement. In typical fashion, Chairman Mamzer decided he would not foot the entire bill himself. Rather, he would steal from a lot of innocent faculty members in the Department of Pediatrics, who had little or no idea of what had happened, to pay for his mistakes using their operating, patient care revenues. In addition, I was informed later, by my friends in the Department, that Chairman Mamzer told its members I had won my case on a minor technicality, and that he was totally innocent. This was but another example of innocent, uninformed people suffering financially and otherwise because of Chairman Mamzer's abusive, immoral and amoral, renegade actions.

1991 - 1994: Director, Maternal-Infant Health and Genetics, Washington State Department of Health; and Clinical Associate Professor, University of Washington School of Public Health and Community Medicine, Seattle

Indeed, there was life after the University of Utah. My almost 10-year, full-time, job with the Washington State Department of Health (WSDOH) from 1991 to 2000 was not unlike my interactions with the UDOH. Working with WSDOH staff and others from all over Washington, we were able to significantly increase access to high quality, comprehensive, community-based, coordinated, culturally competent, and timely maternal and child healthcare services, including genetic services.

When I first started working for WSDOH, I also had a dollar-a-year job as a Clinical Associate Professor at the University of Washington (UW) School of Public Health and Community Medicine (PHCM). The primary objective of this appointment was to coordinate efforts that would improve access to maternal and child healthcare services statewide, with a special emphasis on traditionally underserved populations.

Quantitatively, I had far fewer interactions with physicians when I worked for WSDOH than when I worked at UU. Qualitatively, the interactions were the same! For example, I went to an out-of-state meeting where a group of about 75 medical genetic healthcare providers invited a pediatrician from a large health maintenance organization (HMO) to speak about access to genetic services from an HMO perspective. During the question and answer period, after his presentation, someone in the audience described a newborn with several major birth defects that did not comprise a readily recognizable birth defect syndrome, like Down syndrome. The guest speaker was then asked what would cause him to order a chromosome study on the baby, because chromosome abnormalities are frequently the cause of multiple, major birth defects. He answered, "The amount of money I would get in my end-of-the-year bonus" - and he wasn't joking. The HMO he worked for didn't have a chromosome laboratory and, if he ordered the study, money would have to be sent out of the HMO to pay for the chromosome study, thus causing his year-end bonus to be smaller. He also said his bonus was important to help pay for his children's education. In other words, he would steal money from the funds needed to help his patient and, instead, use it to increase his year-end bonus.

Several years later, I was talking to a physician who was part of a small group practice. He told me about the level of care he was able to give his patients in a non-HMO setting. After several minutes, it became obvious he was seeing some of his patients far more frequently than necessary. I asked him if he knew what it meant for a stockbroker to "churn" his clients. He said no. I told him some brokers made more trades than necessary because they work on a commission basis - just like he did with his patients. I then asked him how he would feel if his broker churned his account. Of course, he acknowledged he wouldn't like it. When I told him he was churning his patients because they were paying on a fee-for-service basis he didn't disagree. However, he said, his patients' health insurance paid for the extra visits. This is another form of stealing, and it sounds a lot like the old joke:

> *Patient: How did my biopsy look, doctor?*
> *Doctor: Well I just talked to my accountant and it looks like you're going to need surgery.*

Obviously, the patient didn't know that s/he had a wallet biopsy, in addition to a tissue biopsy. It's called a conflict of interest, and it has and continues to take place frequently during healthcare evaluations and treatments - most likely including yours!

1994 - 2000: Medical Consultant, Office of Maternal and Child Health, Washington State Department of Health; and 1994 - 2002: Clinical Professor, University of Washington School of Public Health and Community Medicine, Seattle

There was a physician at the UW Medical School who was well known to WSDOH maternal and child health program staff. The physician was working on a project and, at his request, a meeting, including a half dozen health department staff, was set up to discuss his project in order to determine if the WSDOH wanted to fund it.

In preparation for the meeting, WSDOH staff made folders for meeting participants. Each folder contained 8-10 pages of background information. As the meeting progressed, it became more and more obvious things were not going the way this physician wanted. First, he started to talk in a loud voice, and then he started walking around the room while still talking to us. Finally, he tore to shreds all the pages in the folder we had prepared for him, and he threw the confetti in the air. It scattered all over the room. Having dealt with abusive, out of control, loose cannon doctors like him in the past, I was not surprised. After all was said and done, we told him we would not fund his project.

****

A physician from Children's Hospital in Seattle and his co-workers submitted a multi-year, research grant proposal to the Centers for Disease Control and Prevention (CDC). At the same time, WSDOH staff under my direction submitted a somewhat similar proposal to the same CDC office/program. The Children's Hospital physician's proposal aimed to identify, treat, and prevent alcohol-related birth defects from a clinical (patient and family) perspective, while the WSDOH proposal aimed to do the same thing from a public health system-orientation. In fact, the two proposals complimented each other very well. Both proposals were funded and everyone from Children's Hospital, WSDOH, and CDC, agreed the best way to approach the situation was to work together in order to maximize our chances of success, reduce duplication of effort, and minimize disturbing the patients and their families as much as possible.

During the first two years of this collaborative effort, many meetings were held, people were hired to participate in our joint effort, data collection forms were developed, information was collected from a variety of sources, and by mutual agreement all data were to be entered into a computer at Children's Hospital. In the second year of our combined effort, the physician in charge of the Children's Hospital project told me about some computer data entry problems he/they were having. At the same time, more and more data were being collected and about once a month I would ask him how the data entry process was going. He said things were improving, and soon we would be ready to analyze our data.

A few months later we met for lunch. At the meeting, he was supposed to tell me about the progress that was made regarding data entry and analysis, and I was supposed to report on my efforts to get more money to increase the scope of our combined research project.

With a big smile he said two manuscripts about the research had been accepted for publication. I said I had never seen or heard of the manuscripts, and it was inappropriate for him to submit manuscripts for publication without my having reviewed them first. Then he dropped the bomb. He said my name and the names of the others from WSDOH were not in the manuscripts!

We had worked together closely for several years, and until that time we never had a disagreement about anything. Now, however, he was stealing intellectual property, and not for "publish or perish" reasons because he was a Full Professor. I wanted to break both of his legs; however, I didn't for the same reason I didn't throw Dr. Goniff out the window of his 4th floor office at the University of Utah.

Our collaboration ended immediately. There was no way I could work with him again. I could have complained to his chairman, but I believe he would have said to me, "Life's a bitch and then you die. Then they give away your clothes to Goodwill Industries." I also thought of something my <u>first</u> Chairman at the University of Utah told me:

> *Bob, it doesn't make a difference who gets the credit as long as the job gets done.*

Finally, I never openly challenged the physician who "usurped" the WSDOH's fetal alcohol syndrome data because, by that time, I was too busy collecting more information in order to write a book about physicians like him. In addition, I also thought to myself, my father was right. I should have been more careful in dealing with others because in this world there are too many mamzers, kurvahs, and goniffs.

****

The federal government made available maternal and child health grant money aimed at educating doctors and other healthcare providers about clinical/medical genetics. WSDOH staff under my direction applied for one of these multi-year grants. At the same time, a medical geneticist at Children's Hospital in Seattle applied for one. I knew this because there were not that many medical geneticists in Washington State and each of us pretty much knew what the others were doing. In fact, WSDOH had contracts with most of them to help pay for part of their salaries.

I contacted the medical geneticist at Children's Hospital and said I would be most happy to write a strong letter of recommendation for her proposal, and I asked her to do the same for mine. The two grants were important and complimentary. She refused to write a letter of support for my proposal, presumably because her proposal would be in competition with mine for funding at the federal level. What she didn't know was neither of the grant proposals could be submitted to the federal agency without a letter of support from my boss who was the WSDOH maternal and child health director.

With my boss' permission, I told the doctor from Children's Hospital that if she didn't write a letter of support for my grant proposal then my boss would not write a letter of support for her proposal, and her proposal, therefore, would not be fundable. Her response, "Oh, I made a mistake; I'm so sorry; I didn't mean it; I will definitely write a letter of support for your proposal." To make a long story short, both grant proposals were funded and both projects, in the end, were very successful.

Teamwork isn't a common trait among physicians in academic medicine because, unfortunately, many of them can see only the hotdog in their hand and nothing else. Also, too many are driven, insecure, overachievers.

****

The UW School of Public Health and Community Medicine (PHCM) decided to create a public health genetics Master's and PhD degree-granting program. A driving force behind this effort was the UW's hierarchy/central administration. It was making available start-up money for new and innovative programs at the University. This meant the schools, departments, and/or other programs at the University would be in competition with one another for a significant pot of money. I thought the proposed public health genetics program was very important, and I volunteered several hundred hours of my time, without pay, to help the Dean of the School of PHCM lay the initial groundwork for this new enterprise.

A world-renown doctor at the UW <u>Medical School</u> had a particular interest in public health genetics, and he wanted to help the <u>School of PHCM</u> compete for the start-up money. Of course, the UW Medical School had its own pet genetics programs it wanted to get funded. By now you should be able to guess what happened.

At the last minute, the Medical School administration prevailed and its world-renown doctor was forced to remove his name from the School of PHCM proposal because his participation gave the PHCM effort a tremendous amount of credibility. I was at a meeting with the Dean, and others from PHCM and WSDOH, when the Dean found out the world-renown doctor from the Medical School had to be excluded from his public health genetics proposal. He was extremely upset about it.

During the meeting, I suggested the public health core functions made famous in the previously mentioned 1988 Institute of Medicine report be mentioned in our proposal. The Dean exploded. He said the core functions were worthless, they didn't need to be mentioned, and I didn't know what I was talking about. His behavior was rude, boorish, insensitive, and inexcusable.

I wrote him an E-mail the next day telling him I would no longer participate in his effort. I learned from this experience, once again, no good deed goes unpunished. When it came to money, teamwork between the UW Schools of Medicine and PHCM, like programs in other places, didn't exist, even when they shared many common interests. Who said medicine isn't a business that frequently rejects what is best for patients?

Finally, the UW did fund the public health genetics program proposal, and the core functions were included in its educational curriculum - just as I had suggested. By then, the Dean had moved on to a different university where he didn't last very long. This was another example of great recommendations kicking someone upstairs, and to another university.

2002 - Present: Clinical Professor, Department of Pediatrics, University of Washington Medical School, Seattle

I was contacted by a doctor from Texas because I had studied, when I was at the University of Utah, a very large family who had many members affected with a genetic cardiovascular condition associated with significant morbidity and premature death. Over the years, I had kept in contact with one of the matriarchs of the family - a wonderful woman who really cared about the health and well-being for her immediate and extended family, and for humanity, too.

When you read the following letter that I sent to the doctor from Texas, please don't worry about its scientific and technical jargon. Also, when I wrote the letter I was sure something would go wrong with the effort, and I was right.

4 May 2005: A letter I wrote to a physician researcher at the University of Texas Medical School, Houston

> *I am pleased to collaborate with you and your team on your project...*
>
> *Over the past 30+ years, I have worked in the field of medical genetics as a clinician, laboratorian, researcher, teacher, administrator, and mentor. During this time, I authored or co-authored numerous publications and presentations that identify genotype-phenotype correlations for a variety of genetic conditions.*

> *As a participant in this project, I understand my involvement will be to help clinically characterize family members with [DNA] mutations, including the collection of biomarker information such as medical and social history, physical examination, magnetic resonance angiography, biochemical, and other parameters. As someone who has done a great deal of phenotype-genotype correlation work in the past, I believe your project will contribute significantly to the medical literature and benefit many affected individuals and their families. Thus, I would be most happy to collaborate with you and your team in this project.*

The doctor from the University of Texas knew about my long-term interactions with the family. She said she wanted my help to study family members. She asked me to write the above letter to support a National Institutes of Health (NIH) grant proposal she was writing.

I knew her asking for my help was a ruse. She agreed to pay me for my time and effort, and I presumed all she really wanted was an "in" with the family. After her grant was funded, with money included to pay for my services, I never heard from her again. I never expected to because I always thought she was a manure-spreader who was just using me selfishly. I went along for the ride because I knew the experience would make a good vignette to end this chapter.

My amusement turned to real sorrow several months later when I received a phone call from the daughter of the matriarch. Her mom had died. In addition, she told me the doctor from Texas had found out she was a nurse, and that the doctor had asked her to contact other family members to collect medical history information and tissue samples. I told the daughter that was supposed to be my job. The Texas doctor was going to pay her a lot less than what I was supposed to get paid, and according to the daughter the doctor was using guilt-related language to coerce her to do the job. She also said she felt like she was being used like a "prostitute." Could it be the doctor from Texas was, therefore, acting like a pimp?

I told the daughter she should do whatever she thought was best for her and her family, and not to let a doctor from Texas force her to do something she did not want to do.

****

After I observed the enormous financial indiscretions of the University of Utah Department of Pediatrics in the 1980s and 90s, what happened, fifteen years later at the UW Medical School, as described below, didn't surprise me at all.

22 July 2005: "Moving UW forward after billing scandal" *Editorial: Seattle Times* - a daily Seattle newspaper

> *The errors and fraud that led to the University of Washington's $35 million penalty over over-billing government insurance programs should have been avoided.*
>
> *They could have been, according to a report by a special review committee convened to dissect the problems that led to the record fine for a public teaching hospital. The problems were cultural and systemic, the committee reported. In 1996, UW Medical School administrators all but ignored a key federal clarification that services performed by interns and residents can be billed to Medicare and Medicaid only if the teaching physician is present.*
>
> *But administrators, who should have known better, did not heed the warnings of staff members, including one compliance officer who, after he was ignored, filed the federal whistleblower lawsuit that triggered the federal probe.*
>
> *[The Chairman of the Committee] suggested arrogance might have played a role in the disregard of the violations - as well as administrators' deference to department directors who insisted they knew what they were doing.*

> A year ago, this page called for [the Medical School Dean's] resignation because of ample evidence the problems could have been avoided and what seemed to be his minimization of them. Other UW officials also seemed to be in denial. At one point, one UW official said the committee's probe would not consider [the Dean's] performance.
>
> But the committee members clearly went there. Though the committee placed ultimate blame on [the Dean] and urged a thorough performance review, [the Vice Chairman of the Committee] said the panel had full confidence the dean could solve the billing problems. Also, [the President of the University] again expressed confidence in [the Dean], citing the medical school's renowned medical care, research and teaching, and progress he made in correcting billing errors.
>
> The Board of Regents should embrace the committee's recommendations. They are designed to change the culture to one of thorough compliance where employees can express their concerns without fearing for their jobs.
>
> Chief among them is establishment of a full-time position of associate vice president for compliance and risk assessment - something that should have been done long before now.
>
> The committee's review was thorough, frank and, in some instances, scathing. It is a thoughtful roadmap for how the University of Washington can move beyond this disappointing chapter in UW Medicine's otherwise impressive story.

Like I said before, the events that happened at the University of Utah were not isolated incidents. In addition, several of the culprits who were responsible for this debacle at the UW accepted generous severance packages to leave. Now that's what I call being punished for defrauding the patients and taxpayers of our country!

2 February 2006: "Seducing the Medical Profession" *Editorial: New York Times*

> *New evidence keeps emerging that the medical profession has sold its soul in exchange for what can only be described as bribes from the manufacturers of drugs and medical devices. It is long past time for leading medical institutions and professional societies to adopt stronger ground rules to control the noxious influence of industry money on what doctors prescribe for their patients...*

When so many doctors behave out of self-interest, it's no wonder the U.S. Supreme Court ruled in 1982 that medicine is a business and not a profession.

Chapter 6
What Are the Obligations of Physicians, and Why?

> *What's the difference between a real doctor and many of the physicians in the U.S. today? Real doctors have a medical license <u>and</u> they practice in accordance with the oaths and codes of our high calling. The rest are merely pretenders.*

When I was a graduate student/PhD candidate at the Downstate Medical School my thesis mentor and I would play a little game when, every so often, he would ask me, "Bob, what do you know for sure?" I always responded, "If I had something in my hand and I dropped it, it'll definitely hit the floor." Well, forty years have gone by, and I want to tell you about some other things I now know for sure.

My brothers and I never forgot our roots, and I know now what our parents were trying to accomplish when they were raising us. Our father taught us personal responsibility; that is, the obligations and duty we had to ourselves to become honest, hard-working, and productive citizens. Our mother stressed our obligations and our duty to help others. I <u>never</u> heard my parents talk about our rights and, interestingly, the same was true for the Boy Scouts and, especially, its Oath and Law.

Far fewer people in the capitalistic, democratic U.S. would need to be concerned about their rights if children were raised in my parents' home in the Mount Airy section of Philadelphia in the 1950s and 60s. And, finally, in regard to my parents, teachers, and the others from Philadelphia, what I told you about them is not a result of the "halo effect." They really were as I described them in this book.

Regarding my career in medicine, public health, and education, I know for sure the situations/events described in this book are, in microcosm, an accurate reflection of doctors and the medical profession in the U.S. during the past 40 - 50 years, or more. In many ways, unfortunately, doctorhood, medical practice and healthcare in general have been like run-away trains heading in the wrong direction. And, before the predicament gets better, I believe these trains will cause even greater damage than they have done already. Lastly, because a lot of doctors created this mess in the first place, while others stood around watching and doing nothing, it is logical that we doctors are obligated to clean it up.

As an aside, along the way, the run-away healthcare train in the U.S. was taken over, in part, by health insurance and pharmaceutical companies, medical device manufacturers, health maintenance organizations (HMOs), and others. As far as I'm concerned, the term "corporate integrity," as it pertains to our healthcare industry, is an oxymoron until those responsible clean up their acts.

****

Dr. Martin Luther King, Jr., was and still is one of my favorite people and, like him, I have several dreams. The first is one day soon physicians will lead a charge that will dramatically improve the education of doctors and the practice of medicine in our country. I have seen the best and the worst the medical profession has to offer over the past 40+ years and, without a doubt, there is plenty of room for improvement.

The root causes of this situation are a combination of environmental and human/genetic "risk factors." And, make no mistake about it - the outcome of these risk factors is more often than not the infantile behavior that Freud called our "id." Even though doctors and others involved in healthcare are usually book smart and motivated, too many are driven by an infantile pleasure principle that is characterized by amoral and immoral, egocentric, selfish, self-serving, and/or illogical behaviors. And, unfortunately, these immature and sometimes malignant/narcissistic behaviors present themselves frequently as a lust for material gain (greed), control, power, and/or glory.

In addition, there has been and continues to be in too many doctors a "disconnect" between talking-the-talk (the oaths we took) and walking-the-walk (behaving in accordance with the oaths). There are a lot of reasons for this disconnect, and I will describe here only those that I know for sure.

Most medical school admission committees have been much more concerned about students failing medical school than they have been about an applicant's moral compass. Therefore, the grades and test scores of applicants have been considered to be far more important than an applicant's ethics and morality. I have a dream that one day soon medical school admissions committees will consider the moral compass of an applicant as much as they consider his or her "smarts."

There has been and there still is the problem of having the right doctor in the right position at the right time. The knowledge, attitudes, behaviors, and skills needed to be a successful doctor-teacher, researcher, clinician, and/or administrator are not the same. Just because someone is a doctor does not mean he or she is capable of doing all, or even some of these jobs well. I have a dream that the organizations/systems involved in healthcare in our country, and especially at our academic medical centers, which are the last bastions where physicians truly rule, will be much more selective in ensuring that the right doctor is in the right position at the right time.

Doctors have what the majority of patients, families, and communities want and need - an ability, or in some instances only a potential, to improve health, well-being, and longevity. Rather than treating patients and others according to the high standards of our calling, too many doctors treat patients, their families, and others like instruments for the doctor's or the system's well-being. Doctors have patients, their families, and communities over a barrel. To make matters worse, the vast majority of the miscreants who have behaved this way have practiced not only medicine, but also denial. Some of the doctors I have known were the worst cases of "Ostrich Syndrome" I've ever seen.

Being considered a privileged group or class is another important reason why many doctors act the way they do. This attribute was one of the reasons why our physician oaths were created. If ever there was a Catch-22 **[G]** and a cause of aberrant behavior among a significant number of doctors (not to mention a lot of lawyers, dentists, athletes, members of the clergy, elected officials, Hollywood-types, et al.), this is it. There has been too much "me" in the practice of medicine because we doctors and others have not taken to heart:

> *The eternal providence has appointed me to watch over the life and health of Thy creatures. May the love for my art actuate me at all times; may neither avarice nor miserliness, nor thirst for glory or for a great reputation engage my mind; for the enemies of truth and philanthropy could easily deceive me and make me forgetful of my lofty aim of doing good to Thy children.*

A part of the Oath of Moses Maimonides (1135 - 1204 CE) that should be posted in prominent places everywhere doctors work

Too many doctors have been/are more concerned with the "almighty buck" than being holy. I have a dream that one day soon this country's doctors will rise up and live out the true meaning of our physician oaths and creed.

Too many patients, family members, and others have placed doctors on a pedestal, and too many doctors, like so many others, have thought and acted as if they deserved it. In fact, many doctors I have known did as much as they could to perpetuate this myth. Most people don't realize that garbage/trash collectors, and plumbers who install and repair toilets and pipes, have done as much if not more for the public's health than doctors who spend their time seeing patients. I believe this is what George Bernard Shaw meant when he wrote:

> *Democracy means the organization of society for the benefit and at the expense of everybody indiscriminately and not for the benefit of a privileged class.*

If doctors want to emulate someone, let it be the stranger from the story of Joseph in Genesis 37:12 - 18. Jacob sent his son, Joseph, to Shechem to collect information on the status of the family's flock of sheep. When Joseph reached Shechem he came across a man wandering in the fields. The man asked Joseph what he was looking for. Joseph answered: I am looking for my brothers who are tending my family's flock of sheep. The man told Joseph to go to Dothan because that's where his brothers went and, shortly thereafter, that's where Joseph found them.

This unnamed man did a good deed and, at the same time, changed the course of history. No one knows or will ever know his name, or anything else about him. He was not placed on a pedestal, and he didn't become famous, rich, and/or powerful. I have known a lot of doctors like him and, unfortunately, a lot who were exactly the opposite.

****

Regarding individuals who want to be or who are already doctors, and especially for those who aim to be or who are involved already in academic medicine, here are several more things I know for sure:

> *There is a God and you're not Him. And, if you don't believe in God, no problem, just act like you do.*
>
> *There is no promotion, position, or remuneration for doctors that can replace traditional values like integrity and honesty, our families, or performing good deeds.*
>
> *There are a lot of doctors who will steal not only your grants, data, and your money, but also your reputation, and your hopes and dreams.*
>
> *Perform a background check before you work with another doctor and only network or discuss issues/problems with those whom you know and trust.*
>
> *A doctor should not rely on other doctors to help advance his or her career.*

True friendships among doctors in a workplace are rare.

If it is important, get it in writing in order to create the best paper trail possible. Usually this will result in fewer misunderstandings later.

Local, state, and federal rules, regulations, and laws represent the lowest common denominator of human interaction. Therefore, we doctors must strive to interact with patients, families, co-workers, and others at a much higher level.

A smile does not mean yes. Yes does not necessarily mean yes - it might mean yes, no, maybe, we'll see, or I couldn't care less.

Institutional memory rarely exists in agencies, organizations, and institutions where doctors work; for example, in medical schools, clinics, hospitals, HMOs, professional organizations, etc.

Don't be naïve, and do not let doctors and others who behave like mamzers, kurvahs, goniffs, and/or incompetents, grind you down.

Finally, doctors and others who work in our bureaucratic healthcare industry, and especially those who work in academic medical centers, must read *The Prince* by Machiavelli because in it he wrote:

He who wishes to be obeyed must know how to command.

A prince [doctor] *never lacks legitimate reasons to break his* [or her] *promise.*

Princes [including doctors in positions of authority] *and governments are far more dangerous than other elements within society.*

It is not titles [like Doctor, Director, Professor, Chairman, Dean, President, or tenured] *that honor men* [and women], *but men* [and women] *that honor titles.*

Politics [intelligence, social class, and/or a higher education] *have no relation to morals.*

Hatred is gained as much by good works as by evil.

Of mankind [including many doctors] *we may say in general they are fickle, hypocritical, and greedy of gain.*

The fact is that a man [or woman] *who wants to act virtuously in every way necessarily comes to grief among so many who are not virtuous.*

****

For those of you who think I am singling out doctors, I would like to go off on a tangent, briefly, at this time.

In order to improve health and healthcare for patients, families, and communities in the U.S., what we need to do is create and maintain a meaningful healthcare-social contract; that is, an agreement among the members of our society, and between the governed and the government, defining the obligations and duties of each. A meaningful healthcare-social contract is needed, in part, because U.S. Centers for Disease Control and Prevention (CDC) data show that more than half of all premature death and disability in our country is caused by things people do to themselves or to others - for example, smoking, obesity and a lack of exercise, excessive alcohol consumption, drinking and driving, violence aimed at self and/or others, drug abuse, and high-risk sexual behaviors.

I have a dream that one day soon the people in our country will be as passionate about poverty, education, health, healthcare and social services, their jobs, and each other, as they are about their cars, sports, gambling, casual sex, television, movies, social status, and getting drunk. At the same time people are demanding access to appropriate and affordable healthcare and social services, we need to start and/or continue to take better care of ourselves - unless of course there are truly extenuating circumstances.

> *Ask not what your country can do for you - ask what you can do for your country.*
> John F. Kennedy (1917 - 1963)

****

Our country has mandatory, publicly-supported, education systems for <u>all</u> of our youth. Public education serves many different purposes, including the development of a strong, competitive, knowledgeable, and skillful workforce. There is no reason why we shouldn't also have a mandatory, privately and publicly-supported, healthcare system, not to mention food, clothing, shelter, and transportation for <u>all</u> our youth, up to age 18, to compliment the goals and objectives of our education systems.

Unfortunately, the actions of our local, state, and national governments to ameliorate our healthcare and other systemic difficulties have been ineffectual. As part of the solution to this dilemma, I wish I could say we should vote for knowledgeable candidates at every level (local, state, and national) who support an affordable, high quality, universal children's healthcare system, and to vote against those who don't. That's because we have had and we still have in office too many incompetent, self-serving politicians who can't be trusted. The almost 50 million U. S. residents, including a lot of children, who don't have health insurance at this time, prove my point about our elected officials. Chances are the Messiah will appear, or humans will evolve into a higher species, before our elected officials will contribute significantly to solving the problems of our healthcare, social services, and other systems.

****

I am but a simple doctor. Even though I am not trained in political science, philosophy, sociology, history, anthropology, and/or theology, I am sure the best way and, perhaps, the only way for <u>all</u> of us to achieve humankind's greatest aspirations, including peace, justice, happiness, prosperity, and good health, is to always remember and live everyday according to the following eternal truths:

> *If I am not for myself, who will be for me? But if I am for myself alone, what am I? And if not now, when?*
> Hillel the Elder (c. 70 BCE - 10 CE)

> *What is hateful to you, do not do to your fellow man.*
> Hillel the Elder (c. 70 BCE - 10 CE)

and

> *Am I my brother's* [and sister's] *keeper?* [YES!!!]
> Genesis 4:9

Glossary

Bar Mitzvah: literally, in Hebrew, "Son of the Commandment," a Jewish religious ritual commemorating the onset of religious adulthood of a boy at the time of his 13th birthday.

Catch-22: a situation in which a desired outcome or solution is impossible to attain because of a set of inherently illogical rules or conditions; a situation or predicament characterized by absurdity or senselessness; a contradictory or self-defeating course of action (according to www.freedictionary.com).

Character: the combination of qualities or features that distinguishes one person, group, or thing from another; moral or ethical strength; a description of a person's attributes, traits, or abilities.

Clinical Track Physician: salaried faculty members who are committed to excellence in teaching, clinical service, and one or more other areas of professional productivity.

Externship: an unpaid, advanced student or recent graduate program, aimed at assistance in the medical or surgical care of hospital patients, education, training, and/or a research experience.

Galactosemia: a genetic condition caused by inability of the body to metabolize the simple sugar galactose - thereby causing the accumulation of one of its metabolites, galactose 1-phosphate, leading to damage to the liver, central nervous system, and other body systems. Various complications, including death, can occur if an affected individual is not placed on a life-long galactose-free diet beginning shortly after birth.

Metabolic Syndrome: is a group of conditions that places a person at risk for heart disease and diabetes, including high blood pressure, high blood sugar levels, high levels of triglycerides (a type of fat) in the blood, low levels of high-density lipoprotein (the good cholesterol) in the blood, too much fat in the waist area.

Phenylketonuria (PKU): a genetic condition in which the body does not properly metabolize the amino acid phenylalanine. Severe mental retardation occurs if an affected individual is not placed on a very low, life-long phenylalanine diet beginning shortly after birth.

Principal Investigator (PI): the individual who is assigned the primary responsibility for the proper conduct and management of the project, including its technical, programmatic, logistical, physical, financial, regulatory compliance, legal, and ethical aspects.

Profession: a disciplined group of individuals who adhere to high ethical standards and uphold themselves to, and are accepted by, the public as possessing special knowledge and skills in a widely recognized, organized body of learning derived from education and training at a high level, and who are prepared to exercise this knowledge and these skills in the interest of others. Inherent in this definition is the concept that the responsibility for the welfare, health and safety of the community shall take precedence over other considerations (according to the Australian Competition and Consumer Commission).

Secondary Research Track Appointment: a non-tenured, non-salaried faculty member whose primary professional efforts are devoted to one or more research projects.

Tenure: accurately and unequivocally defined, lays no claim whatever to a guarantee of lifetime employment. Rather, tenure provides only that no person continuously retained as a full-time faculty member beyond a specified lengthy period of probationary service may thereafter be dismissed without adequate cause (according to the American Association of University Professors).

Tenure Track Physician: salaried faculty who are committed to excellence in teaching, scholarship (research), and service.

Yiddish: a language of the Ashkenazi Jews of Central and Eastern Europe, resulting from a fusion of medieval German dialects, and secondarily from Hebrew, Aramaic, various Slavic languages, Old French, and Old Italian.

Appendix A: Physicians' Prayer, Oaths, and Codes

Daily Prayer of a Physician

Attributed to Moses Maimonides, a 12th century Jewish physician in Egypt, but probably written by Marcus Herz, a German physician and pupil of Immanuel Kant. It first appeared in 1793:

Almighty God, Thou has created the human body with infinite wisdom. Ten thousand times ten thousand organs hast Thou combined in it that act unceasingly and harmoniously to preserve the whole in all its beauty the body which is the envelope of the immortal soul. They are ever acting in perfect order, agreement and accord. Yet, when the frailty of matter or the unbridling of passions deranges this order or interrupts this accord, then forces clash and the body crumbles into the primal dust from which it came. Thou sendest to man diseases as beneficent messengers to foretell approaching danger and to urge him to avert it.

Thou has blest Thine earth, Thy rivers and Thy mountains with healing substances; they enable Thy creatures to alleviate their sufferings and to heal their illnesses. Thou hast endowed man with the wisdom to relieve the suffering of his brother, to recognize his disorders, to extract the healing substances, to discover their powers and to prepare and to apply them to suit every ill. In Thine Eternal Providence Thou hast chosen me to watch over the life and health of Thy creatures. I am now about to apply myself to the duties of my profession. Support me, Almighty God, in these great labors that they may benefit mankind, for without Thy help not even the least thing will succeed.

Inspire me with love for my art and for Thy creatures. Do not allow thirst for profit, ambition for renown and admiration, to interfere with my profession, for these are the enemies of truth and of love for mankind and they can lead astray in the great task of attending to the welfare of Thy creatures. Preserve the strength of my body and of my soul that they ever be ready to cheerfully help and support rich and poor, good and bad, enemy as well as friend. In the sufferer let me see only the human being. Illumine my mind that it recognize what presents itself and that it may comprehend what is absent or hidden. Let it not fail to see what is visible, but do not permit it to arrogate to itself the power to see what cannot be seen, for delicate and indefinite are the bounds of the great art of caring for the lives and health of Thy creatures. Let me never be absent-minded. May no strange thoughts divert my attention at the bedside of the sick, or disturb my mind in its silent labors, for great and sacred are the thoughtful deliberations required to preserve the lives and health of Thy creatures.

Grant that my patients have confidence in me and my art and follow my directions and my counsel. Remove from their midst all charlatans and the whole host of officious relatives and know-all nurses, cruel people who arrogantly frustrate the wisest purposes of our art and often lead Thy creatures to their death.

Should those who are wiser than I wish to improve and instruct me, let my soul gratefully follow their guidance; for vast is the extent of our art. Should conceited fools, however, censure me, then let love for my profession steel me against them, so that I remain steadfast without regard for age, for reputation, or for honor, because surrender would bring to Thy creatures sickness and death.

Imbue my soul with gentleness and calmness when older colleagues, proud of their age, wish to displace me or to scorn me or disdainfully to teach me. May even this be of advantage to me, for they know many things of which I am ignorant, but let not their arrogance give me pain. For they are old and old age is not master of the passions. I also hope to attain old age upon this earth, before Thee, Almighty God!

Let me be contented in everything except in the great science of my profession. Never allow the thought to arise in me that I have attained to sufficient knowledge, but vouchsafe to me the strength, the leisure and the ambition ever to extend my knowledge. For the art is great, but the mind of man is ever expanding.

Almighty God! Thou hast chosen me in Thy mercy to watch over the life and death of Thy creatures. I now apply myself to my profession. Support me in this great task so that it may benefit mankind, for without Thy help not even the least thing will succeed.

Translated by Harry Friedenwald, Bulletin of the Johns Hopkins Hospital, 1917

Physician's Oath of the World Medical Association

At the time of being admitted as a member of the medical profession:

I solemnly pledge myself to consecrate my life to the service of humanity;

I will give to my teachers the respect and gratitude which is their due;

I will practice my profession with conscience and dignity; the health of my patient will be my first consideration;

I will maintain by all the means in my power, the honor and the noble traditions of the medical profession; my colleagues will be my brothers;

I will not permit considerations of religion, nationality, race, party politics or social standing to intervene between my duty and my patient;

I will maintain the utmost respect for human life from the time of conception, even under threat, I will not use my medical knowledge contrary to the laws of humanity;

I make these promises solemnly, freely and upon my honor.

Weill Cornell Medical College's Hippocratic Oath

I do solemnly vow, to that which I value and hold most dear:

That I will honor the Profession of Medicine, be just and generous to its members, and help sustain them in their service to humanity;

That just as I have learned from those who preceded me, so will I instruct those who follow me in the science and the art of medicine;

That I will recognize the limits of my knowledge and pursue lifelong learning to better care for the sick and to prevent illness;

That I will seek the counsel of others when they are more expert so as to fulfill my obligation to those who are entrusted to my care;

That I will not withdraw from my patients in their time of need;

That I will lead my life and practice my art with integrity and honor, using my power wisely;

That whatsoever I shall see or hear of the lives of my patients that is not fitting to be spoken, I will keep in confidence;

That into whatever house I shall enter, it shall be for the good of the sick;

That I will maintain this sacred trust, holding myself far aloof from wrong, from corrupting, from the tempting of others to vice;

That above all else I will serve the highest interests of my patients through the practice of my science and my art;

That I will be an advocate for patients in need and strive for justice in the care of the sick.

I now turn to my calling, promising to preserve its finest traditions, with the reward of a long experience in the joy of healing.

I make this vow freely and upon my honor.

****

Oath of the Imperial College of the United Kingdom School of Medicine

Now, as a new doctor, I solemnly promise that I will to the best of my ability serve humanity—caring for the sick, promoting good health, and alleviating pain and suffering.

I recognize that the practice of medicine is a privilege with which comes considerable responsibility and I will not abuse my position.

I will practice medicine with integrity, humility, honesty, and compassion—working with my fellow doctors and other colleagues to meet the needs of my patients.

I shall never intentionally do or administer anything to the overall harm of my patients.

I will not permit considerations of gender, race, religion, political affiliation, sexual orientation, nationality, or social standing to influence my duty of care.

I will oppose policies in breach of human rights and will not participate in them. I will strive to change laws that are contrary to my profession's ethics and will work towards a fairer distribution of health resources.

I will assist my patients to make informed decisions that coincide with their own values and beliefs and will uphold patient confidentiality.

I will recognize the limits of my knowledge and seek to maintain and increase my understanding and skills throughout my professional life. I will acknowledge and try to remedy my own mistakes and honestly assess and respond to those of others.

I will seek to promote the advancement of medical knowledge through teaching and research.

I make this declaration solemnly, freely, and upon my honor.

****

American Medical Association Principles of Medical Ethics

A physician shall be dedicated to providing competent medical care, with compassion and respect for human dignity and rights.

A physician shall uphold the standards of professionalism, be honest in all professional interactions, and strive to report physicians deficient in character or competence, or engaging in fraud or deception, to appropriate entities.

A physician shall respect the law and also recognize a responsibility to seek changes in those requirements which are contrary to the best interests of the patient.

A physician shall respect the rights of patients, colleagues, and other health professionals, and shall safeguard patient confidences and privacy within the constraints of the law.

A physician shall continue to study, apply, and advance scientific knowledge, maintain a commitment to medical education, make relevant information available to patients, colleagues, and the public, obtain consultation, and use the talents of other health professionals when indicated.

A physician shall, in the provision of appropriate patient care, except in emergencies, be free to choose whom to serve, with whom to associate, and the environment in which to provide medical care.

A physician shall recognize a responsibility to participate in activities contributing to the improvement of the community and the betterment of public health.

A physician shall, while caring for a patient, regard responsibility to the patient as paramount.

A physician shall support access to medical care for all people.

****

Australian Medical Association Code of Ethics

<u>*The Doctor and the Patient: Patient Care*</u>

Consider first the well-being of your patient.

Treat your patient with compassion and respect.

Approach healthcare as a collaboration between doctor and patient.

Practice the science and art of medicine to the best of your ability.

Continue lifelong self-education to improve your standard of medical care.

Maintain accurate contemporaneous clinical records.

Ensure that doctors and other health professionals upon whom you call to assist in the care of your patients are appropriately qualified.

Make sure that you do not exploit your patient for any reason.

Avoid engaging in sexual activity with your patient.

Refrain from denying treatment to your patient because of a judgment based on discrimination.

Respect your patient's right to choose their doctor freely, to accept or reject advice and to make their own decisions about treatment or procedures.

Maintain your patient's confidentiality. Exceptions to this must be taken very seriously. They may include where there is a serious risk to the patient or another person, where required by law, where part of approved research, or where there are overwhelming societal interests.

Upon request by your patient, make available to another doctor a report of your findings and treatment.

Recognize that an established therapeutic relationship between doctor and patient must be respected.

Having initiated care in an emergency setting, continue to provide that care until your services are no longer required.

When a personal moral judgment or religious belief alone prevents you from recommending some form of therapy, inform your patient so that they may seek care elsewhere.

Recognize that you may decline to enter into a therapeutic relationship where an alternative healthcare provider is available, and the situation is not an emergency one.

Recognize that you may decline to continue a therapeutic relationship. Under such circumstances, you can discontinue the relationship only if an alternative healthcare provider is available and the situation is not an emergency one. You must inform your patient so that they may seek care elsewhere.

Recognize your professional limitations and be prepared to refer as appropriate.

Place an appropriate value on your services when determining any fee. Consider the time, skill, and experience involved in the performance of those services together with any special circumstances.

Ensure that your patient is aware of your fees where possible. Encourage open discussion of healthcare costs.

When referring your patient to institutions or services in which you have a direct financial interest, provide full disclosure of such interest.

If you work in a practice or institution, place your professional duties and responsibilities to your patients above the commercial interests of the owners or others who work within these practices.

Ensure security of storage, access and utilization of patient information.

Protect the right of doctors to prescribe, and any patient to receive, any new treatment, the demonstrated safety and efficacy of which offer hope of saving life, re-establishing health or alleviating suffering. In all such cases, fully inform the patient about the treatment, including the new or unorthodox nature of the treatment, where applicable.

<u>The Doctor and the Patient: Clinical Research</u>

Accept responsibility to advance medical progress by participating in properly developed research involving human participants.

Ensure that responsible human research committees appraise the scientific merit and the ethical implications of the research.

Recognize that considerations relating to the well-being of individual participants in research take precedence over the interests of science or society.

Make sure that all research participants or their agents are fully informed and have consented to participate in the study. Refrain from using coercion or unconscionable inducements as a means of obtaining consent.

Inform treating doctors of the involvement of their patients in any research project, the nature of the project and its ethical basis.

Respect the participant's right to withdraw from a study at any time without prejudice to medical treatment.

Make sure that the patient's decision not to participate in a study does not compromise the doctor-patient relationship or appropriate treatment and care.

Ensure that research results are reviewed by an appropriate peer group before public release.

The Doctor and the Patient: Clinical Teaching

Honor your obligation to pass on your professional knowledge and skills to colleagues and students.

Before embarking on any clinical teaching involving patients, ensure that patients are fully informed and have consented to participate.

Respect the patient's right to refuse or withdraw from participating in clinical teaching at any time without compromising the doctor-patient relationship or appropriate treatment and care.

Avoid compromising patient care in any teaching exercise. Ensure that your patient is managed according to the best-proven diagnostic and therapeutic methods and that your patient's comfort and dignity are maintained at all times.

Where relevant to clinical care, ensure that it is the treating doctor who imparts feedback to the patient.

Refrain from exploiting students or colleagues under your supervision in any way.

The Doctor and the Patient: The Dying Patient

Remember the obligation to preserve life, but, where death is deemed to be imminent and where curative or life-prolonging treatment appears to be futile, try to ensure that death occurs with dignity and comfort.

Respect the patient's autonomy regarding the management of their medical condition including the refusal of treatment.

Respect the right of a severely and terminally ill patient to receive treatment for pain and suffering, even when such therapy may shorten a patient's life.

Recognize the need for physical, psychological, emotional, and spiritual support for the patient, the family and other carers not only during the life of the patient, but also after their death.

The Doctor and the Patient: Transplantation

Recognize that a potential donor is entitled to the same standard of care as any other patient.

Inform the donor and family fully of the proposal to transplant organs, the purpose and the risks of the procedure.

Exercise sensitivity and compassion when discussing the option to donate organs with the potential donor and family.

Refrain from using coercion when obtaining consent to all organ donations.

Explain brain death to potential donor families. Similarly explain that continued artificial organ support is necessary to enable subsequent organ transplantation.

Ensure that the determination of the death of any donor is made by doctors who are neither involved with the transplant procedure nor caring for the proposed recipient.

Recognize the important contribution donor families make in difficult circumstances. Ensure that they are given the opportunity to receive counseling and support.

The Doctor and the Profession: Professional Conduct

Build a professional reputation based on integrity and ability.

Recognize that your personal conduct may affect your reputation and that of your profession.

Refrain from making comments which may needlessly damage the reputation of a colleague.

Report suspected unethical or unprofessional conduct by a colleague to the appropriate peer review body.

Where a patient alleges unethical or unprofessional conduct by another doctor, respect the patient's right to complain and assist them in resolving the issue.

Accept responsibility for your psychological and physical well-being as it may affect your professional ability.

Keep yourself up to date on relevant medical knowledge, codes of practice and legal responsibilities.

The Doctor and the Profession: Advertising

Confine advertising of professional services to the presentation of information reasonably needed by patients or colleagues to make an informed decision about the availability and appropriateness of your medical services.

Make sure that any announcement or advertisement directed towards patients or colleagues is demonstrably true in all respects. Advertising should not bring the profession into disrepute.

Do not publicly endorse therapeutic goods as defined under the Therapeutic Goods Act 1989, contrary to the Therapeutic Goods Advertising Code.

Exercise caution in publicly endorsing any particular commercial product or service not covered by the Therapeutic Goods Advertising Code.

Ensure that any therapeutic or diagnostic advance is described and examined through professional channels, and, if proven beneficial, is made available to the profession at large.

<u>The Doctor and the Profession: Referral to Colleagues</u>

Obtain the opinion of an appropriate colleague acceptable to your patient if diagnosis or treatment is difficult or obscure, or in response to a reasonable request by your patient.

When referring a patient, make available to your colleague, with the patient's knowledge and consent, all relevant information and indicate whether or not they are to assume the continuing care of your patient during their illness.

When an opinion has been requested by a colleague, report in detail your findings and recommendations to that doctor.

Should a consultant or specialist find a condition which requires referral of the patient to a consultant in another field, only make the referral following discussion with the patient's general practitioner – except in an emergency situation.

Professional Independence

In order to provide high quality healthcare, you must safeguard clinical independence and professional integrity from increased demands from society, third parties, individual patients and governments.

Protect clinical independence as it is essential when choosing the best treatment for patients and defending their health needs against all who would deny or restrict necessary care.

Refrain from entering into any contract with a colleague or organization which may conflict with professional integrity, clinical independence or your primary obligation to the patient.

Recognize your right to refuse to carry out services which you consider to be professionally unethical, against your moral convictions, imposed on you for either administrative reasons or for financial gain or which you consider are not in the best interest of the patient.

The Doctor and Society

Endeavour to improve the standards and quality of, and access to, medical services in the community.

Accept a share of the profession's responsibility to society in matters relating to the health and safety of the public, health education and legislation affecting the health of the community.

Use your special knowledge and skills to minimize wastage of resources, but remember that your primary duty is to provide your patient with the best available care.

Make available your special knowledge and skills to assist those responsible for allocating healthcare resources.

Recognize your responsibility to give expert evidence to assist the courts or tribunals.

When providing scientific information to the public, recognize a responsibility to give the generally held opinions of the profession in a form that is readily understood. When presenting any personal opinion which is contrary to the generally held opinion of the profession, indicate that this is the case.

Regardless of society's attitudes, ensure that you do not countenance, condone or participate in the practice of torture or other forms of cruel, inhuman, or degrading procedures, whatever the offence of which the victim of such procedures is suspected, accused or convicted.

Appendix B: Reasons Why We Became Physicians at the End of the 20th Century in the United States

The following list is not meant to be exhaustive, nor is it in any particular order:

My parents wanted me to be one

To have a better chance of evading military service when there was a draft

To achieve status, recognition, respect, glory, and/or admiration

To achieve material gain and/or a higher standard of living

To achieve power, influence, and/or control over other people

To achieve autonomy

To achieve long-term security

To specialize to the nth degree, or to wear lots of different "hats," or to be some place in between

To achieve job mobility; that is, to be able to switch jobs with little or no extra effort

To be creative, innovative, challenged intellectually, and/or to learn new things every day

To be a teacher and/or a mentor

To affiliate and/or develop friendships with like-minded people

To help not only patients and families, but whole communities and populations in ways involving their physical and mental health and well-being

To search for ideas and ideals that are valuable for their own sake

When I was a member of the UU Medical School Admissions Committee, student applicants mentioned, during their interviews, most of these reasons for wanting to become a physician.

Appendix C: A *Deseret News* Newspaper Article and Two University of Utah Committee Reports

Document 1: 9 August 1991, "Two Doctors File Bias Suits Against U" *The Deseret News*

Two doctors who have filed lawsuits against the University of Utah School of Medicine say the school practices more than medicine.

According to the doctors, it practices discrimination and retaliation as well.

A U doctor in the Department of Pediatrics and a former radiology resident have filed separate suits in District Court accusing the school of discriminating against them on the basis of age, sex and race. Each doctor says the school also retaliated when they attempted to battle the discrimination...

In an April 5 suit, Dr. Patrick Bray claims medical school officials [including Chairman Mamzer according to a discussion I had with Dr. Bray] asked him to resign in 1987 as director of his division because of his age. Dr. Bray was the founder of the division [of Pediatric Neurology]. When he did not resign, the U cut his salary by $10,000 per year, Dr. Bray says in his suit.

Also beginning in 1987, the U required Dr. Bray to subsidize his salary by working several days a week as a clinician at the Children's hospital. At the same time, the U denied Dr. Bray the only sabbatical he requested during his 27 years at the medical school "in spite of the fact that other,

> *younger faculty members request and are regularly granted sabbaticals," the suit says.*
>
> *In a letter, a U official told the 64-year-old Dr. Bray the time had come for him to retire. Medical school officials slashed his teaching responsibilities, removed students from his clinics and gave his laboratory space to a new, younger faculty member, Dr. Bray says in his suit...*
>
> *Dr. Bray seeks $40,000 in back wages stemming from his 1987 pay cut. If a jury concludes that the medical school violated federal discrimination laws, Dr. Bray is entitled to double damages totaling $80,000, said Dr. Bray's attorney...*

Dr. Bray was a very knowledgeable, hard-working, and honorable person. That was probably why he was treated the way he was by the UU hierarchy. Unfortunately, I lost contact with Dr. Bray and was unable to find out the outcome his lawsuit.

Document 2: 4 December 1989, *From the Chair of the UU Retention, Promotion, and Tenure (RPT) Standards and Appeals Committee to the President of the University: Report of the Appeal of Dr. Robert Fineman*

> *This is a complex case. Dr. Fineman is involved in a number of serious disputes with other members of the Department of Pediatrics, and the file makes it clear that a good deal of animosity exists between Dr. Fineman, the Department Chair [Chairman Mamzer], and the Division of Medical Genetics Chief [Dr. Kurveh]. It is difficult to know how much these conflicts and other matters inappropriate to promotion decisions influenced the outcome.*
>
> *Recognition of these uncertainties appears to be the reason for the unusual actions of the University Promotion and Tenure Advisory*

Committee (UPTAC), i.e., its two votes and the overwhelming number of abstentions. Based on UPTAC's actions, the Vice President for Health Sciences requested that the Dean of the School of Medicine initiate another review of the file, although no provision for such a procedure is specified in Faculty Regulations. The request was rejected, and therefore the file was reviewed by the Vice President for Health Sciences and forwarded to the Provost with a recommendation against promotion.

The RPT Standards and Appeals Committee does not believe the Provost's recommendation against promotion was based on grounds that are arbitrary, capricious, or discriminatory. Dr. Fineman's charges, however, do not focus primarily on these issues but, rather, on the claim that the Provost's recommendation was unreasonable in view of fundamentally flawed procedures in the review process which denied the appellant, Dr. Fineman, basic fairness and due process. **The Committee by a unanimous vote (10 - 0) finds that the review process was in fact deeply flawed and heavily biased against Dr. Fineman. We take no stand on whether or not he deserves promotion on the basis of his merits with regard to research, teaching and service, but we conclude that his review was neither fair nor served due process [emphasis mine - note also that it says research, teaching, and service].**

The Committee finds that the following actions clearly reveal serious procedural errors which biased the review:

Letters of evaluation were solicited from individuals within the University where documented evidence of interpersonal conflicts between the appellant and the writer of one or more letters makes it highly unlikely that the writers could provide a fair and balanced evaluation.

A [negative] letter from [Dr. Goniff] was not provided to the Department of Pediatrics Promotion, Retention and Tenure (PRT) Committee, but was added to the file subsequently. This violates Faculty Regulations which specifically requires that the materials shall be submitted to the Department PRT Committee and makes no provisions for the addition of materials subsequent to their review other than recommendations from other appropriate parties and the candidate's response to these recommendations.

A favorable letter of evaluation from the Director of the Handicapped Children's Services, Utah State Department of Health, was not included in the review file, despite the fact that it was received in time for the review process. [In fact, two favorable evaluation letters never made it into my file at the Department level - the second one was from a Professor from the Harvard Medical School.] Although failure to include this letter is attributed to clerical error, it indicates the absence of care in preparing and submitting a complete file for review.

The minutes of the Department of Pediatrics PRT Committee contained no evidence that an effort was made to address Dr. Fineman's professional criteria in which a faculty member in the Department is to be judged for consideration to promotion to the rank of Professor.

Information contained in a written response to this Committee from the Chair of the Department of Pediatrics PRT Committee, as well as the unusually large number of letters solicited, provide evidence that the Chairman of the Department of Pediatrics' selection of external reviewers was designed to offset individuals recommended by the appellant. [Chairman Mamzer's] selections, however, should have aimed for individuals who would have provided reasoned, informed recommendations with regard to the professional qualifications of the candidate.

The soliciting of external letters by [Chairman Mamzer] was seriously tainted by the addition of handwritten notes appended to some unknown number of letters addressed to external referees. On one letter the note says, "I am sorry we did not get to talk on 9/8 and 9/9. I hope you can help me with this promotion. We need real rigor since it is likely to be lacking internally." The inference of this note asking for a negative recommendation is unmistakable.

[A friend gave me a copy of one of Chairman Mamzer's "seriously tainted" letters addressed to an external evaluator. After I obtained it, I contacted a chairman I knew from another medical school. I told him about Chairman Mamzer's handwritten note, and I asked him if it was commonplace for Chairs to influence an evaluator's letter. He said, "Of course, it's done all the time. Evaluators are told to write good things when a Chair wants a candidate to be promoted and bad things when he or she doesn't." I said to him, "But that could make the promotion process nothing more than a popularity contest." He said, "That's right! "]

[Chairman Mamzer] in his [evaluation] *letter failed to provide an adequate evaluation of the applicant's qualifications with regard to research, teaching and service. Faculty Regulations require that the Chair include specific reasons for his recommendation. The Chair's letter, however, deals largely with matters extraneous to the criteria spelled out for consideration for promotion to rank of Professor in the Department. It refers*

primarily to personality issues and personal characteristics of the appellant with little regard to his professional contributions in specific areas of evaluation. Furthermore, it includes direct criticism of the candidate for exercising his right to the solicitation of letters on a non-confidential basis, a right guaranteed in Faculty Regulations. While the reporting of the candidate's decision to exercise that right is appropriate, the Chair's comments on that exercise are, in our judgment, inappropriate.

[Chairman Mamzer] failed to deliver a copy of his [negative] recommendation in a timely fashion to the candidate, thereby denying him his right to respond to that recommendation prior to consideration by higher levels of review. Given the nature of this case, the Chair should have been careful to ensure that the candidate received the recommendation on time.

The Department of Pediatrics uses a procedure which allows all of its members to participate in the review process and vote, whether eligible to vote or not. In our judgment, the inclusion in the file of the informal vote by non-eligible members could only serve to influence, inappropriately, the decisions made at higher levels of review.

The Dean of the School of Medicine failed to provide a timely review of the request for promotion, delaying the actual review by three months. Such a delay is unwarranted, resulting in UPTAC receiving the file late in the year. Consequently, the Vice President for Health Science's request for consideration could not be implemented in a timely manner, resulting in the denial of the re-review of a substantially flawed file.

The Committee unanimously finds that these matters represent substantial defects that

> ***denied the appellant fairness and due process* [emphasis mine].** *The Committee, however, is divided as to the appropriate remedy. A minority of three members believe that, in light of what has happened so far, it would be impossible to secure a fair review of Dr. Fineman in the years to come. These members feel strongly that the review process was inherently unfair and that a new review would be extremely inhumane. Consequently, they recommend that the President of the University grant promotion to the rank of Professor at this time.*
>
> *The majority of seven, on the other hand, believes that professional, balanced judgments of the record and performance, consistent with Department, School of Medicine, and University PRT criteria, must serve as the basis of a positive recommendation for tenure or promotion. Consequently, they recommend that the President require a completely new review under the direct supervision of the Vice President for Health Sciences, who shall be responsible for insuring that the review is conducted in a manner consistent with established policies and devoid of any violations of procedures. They also recommend that if, in the President's judgment, a fair review is no longer possible, the President should recommend promotion forthwith.*

Document 3: 12 March 1990, *From the Chair of the UU Academic Freedom and Tenure Committee (AFTC) to the President of the University: Report and Recommendations of the AFTC on the Complaint of Dr. Robert Fineman*

> *In May 1988 Dr. Robert Fineman initiated an informal inquiry to the AFTC concerning certain aspects of his treatment by [Dr. Mamzer], his Chairman, and his Division Chief, [Dr. Kurveh]. This inquiry resulted in a formal letter of complaint*

against Drs. Mamzer and Kurveh (29 June 1988). In an effort to understand and, where appropriate, help resolve the issues raised in the complaint, the AFTC authorized [last year's Chair of the Committee] to make inquiries and suggest solutions. This effort resulted in a Committee recommendation to the parties on 14 December 1988 that they engage in arbitration of their differences. When the parties failed to take that action, the Committee made a report to you on 17 January 1989, again recommending arbitration. [In fact, I accepted the Committee's recommendation for arbitration - Drs. Mamzer and Kurveh didn't; and the Dean of the Medical School, the Vice President for Health Sciences, and the President of the University agreed with Drs. Mamzer and Kurveh]. On 19 May 1989, you wrote to the parties treating the matter as closed. Meanwhile, Dr. Fineman continued to make informal complaints about further alleged objectionable treatment of him by [Drs. Kurveh and Mamzer].

On 21 September 1989, Dr. Fineman asked the Committee to hold a hearing into his allegations in order to bring the matter to closure. The Committee voted on 4 October 1989 to hold a formal hearing. The hearing was scheduled in two parts. First an initial hearing would inquire whether there were violations of academic freedom or tenure rights involved if the actions alleged by Dr. Fineman were found to have occurred. If the allegations were determined to state claims of violation, then the second part of the hearing would be held to make findings of facts.

On 4 December 1989, the first part of the hearing was held. The Committee determined that a potential violation of academic freedom and tenure rights existed in three areas.

The Committee scheduled the second part of the hearing into these areas for 2 February 1990. Both parties were given until 12 January 1990 to submit written information and until 18 January 1990 to respond to the other side's submission. Nothing was received from [Drs. Mamzer and Kurveh] until 26 January 1990, one week before the scheduled hearing. The Committee therefore conducted the hearing without the benefit of prior responsive submissions.

The Committee carefully examined the documentation provided it and heard six hours of testimony presented [by the two sides] on 2 February 1990. The Committee then met to reach its decision on the three points. What follows are its findings, recommended remedies, and the reasoning which led to these conclusions.

I. Retaliation

Finding
The Committee finds that [Drs. Kurveh and Mamzer] acted in retaliation to Dr. Fineman's filing a complaint with the AFTC (29 June 1988) in demoting him from Director of the chromosome laboratory to Acting Director. The vote was - Yes: 8; No: 0; Abstain: 1.

Remedy
The AFTC recommends that Dr. Fineman be reinstated immediately as Director of the chromosome laboratory. His performance in that capacity should be subject to review in six months (or sooner should it be decided to initiate a "show cause" hearing to remove him from the position). This review should be conducted under the direction of the Vice President for Health Sciences and performed by an impartial committee of external reviewers selected from a list compiled by the Vice President as a result of suggestions made by Drs. Fineman, [Mamzer,

and Kurveh]. The vote was - Yes: 9; No: 0; Abstain: 0.

<u>Reasons</u>
It would be difficult ever to find direct evidence of a retaliatory motive in this kind of case. No sensible administrator would say, "I am doing this to you because of the complaint you filed." Therefore, the Committee deemed it necessary to break the retaliatory question into two questions. First, were the actions causally linked to the filing of the original complaint so that a <u>prima facie</u> case of retaliation can be made? Second, is there satisfactory evidence of an alternative explanation for these actions other than retaliation? These questions necessarily rely heavily on the timing and justifications offered in support of the actions.

In the letter of 14 December 1988, [Dr. Kurveh] stated that his opinions had changed as a result of "having to review all of our issues of difference in the letters for the Academic Freedom Committee" and ensuing discussions; that he had "come to feel differently about these issues" than he had before; and that it was impossible to "return to the place where we were prior to this summer." This letter, even in the context of an ongoing strained relationship, establishes a causal link between the filing of the complaint and actions taken, thus making a <u>prima facie</u> case for retaliation. The question then is whether there is a sufficient alternative explanation.

[Drs. Mamzer and Kurveh] replied that the timing was coincidental and that they had long had concerns about Dr. Fineman's performance in the laboratory. They assert that he was not "hands on" enough and that "Dr. Fineman was not doing his job." But the written evidence of dissatisfaction with Dr. Fineman's performance in the chromosome laboratory is most *unsatisfactory. Moreover, clear opportunities for commenting on Dr. Fineman's performance were bypassed. In his* [recent] *recommendation against Dr. Fineman's promotion* [from Associate Professor to Full Professor], *written two months before the* [laboratory directorship] *demotion, [Dr. Kurveh] makes no mention of unsatisfactory performance in the lab. Surely in a letter designed to justify non-promotion, the author could be expected to mention every negative factor, yet the 12 October 1988 letter actually praises Dr. Fineman's performance in the lab. And in light of the length of his service in this position, if administrators were dissatisfied with his performance, normal expectations would be that he would be contacted and allowed to discuss the matter before any actions were taken. Corroborating evidence can be found in [Dr. Mamzer's] extremely negative attitude throughout the Committee process. In response to the AFTC on 26 January 1990, he characterizes Dr. Fineman's appeal as "methods of manipulating his supervisors and his colleagues into doing things his way." In an earlier letter to [the VP for Health Sciences] of 18 May 1989, [Dr. Mamzer] asked that the Committee be "censured" for its efforts to resolve the dispute informally. When asked directly by the Committee in its letter of 13 December 1989 to provide an alternative explanation of these actions and/or to counter Dr. Fineman's claims of retaliation, [Dr. Mamzer] said only that he "cannot tell from the Committee's letter the precise information which you are requesting from me" (letter to the Committee dated 26 January 1990). It is in this same letter that he characterizes appeals to the Committee and the University Retention, Promotion, and Tenure (RPT) Standards and Appeals Committee as "methods of manipulating his supervisors and colleagues into doing things his way." These comments betray an attitude toward university processes that support the inference of retaliation for exercise of rights provided by those processes.*

In reaching this conclusion, the Committee does not intend to imply that [Drs. Kurveh and Mamzer] acted from malicious motivations. We mean only that Dr. Fineman's complaining of what he perceived to be illegitimate treatment resulted in his being subjected to punitive actions on the part of his Department Chair and Division Chief regardless of their thoughts about their actions. It is essential that faculty not only be told that they may appeal to the Committee in such cases, but that the University focus all its efforts on removing any practical obstacles to a faculty member's actually exercising this right. When such obstacles occur in the form of retaliation on the part of one's superiors, it is essential both that the University takes action to rectify the situation and that it takes steps to ensure that such illegitimate behavior does not occur in the future.

II. Private Practice Income (PPI)

<u>Finding</u>
The Committee finds that Dr. Fineman's rights have been violated in two ways: 1) by changing the basis for the calculation of his PPI in the middle of the fiscal year in which he already negotiated a right, and 2) by treating him on a patient billing format disparately from anyone else in the Division of Medical Genetics. The vote was - Yes: 8; No: 0; Abstain: 0.

<u>Remedy</u>
The Committee recommends that Dr. Fineman be paid for the fiscal years 1988 - 1989 and 1990 - 1991 at the previously negotiated rate for 1988 - 1989, minus any across the board cuts that have been taken by everyone in the Division of Medical Genetics for those two years. The vote was - Yes: 8; No: 0; Abstain: 0.

Reasons

[Drs. Kurveh and Mamzer] agreed that the basis for calculating Dr. Fineman's PPI was changed in the middle of the [academic] year [July 1st - June 30th], and that the change made him unique within the Division. They argued that this was not an issue relevant to academic freedom or tenure rights. In the sense that tenure rights have to do with security of job conditions for academic personnel, this is a tenure right. Moreover, this Committee is designated by University Regulations as the site for general grievances by faculty. Regardless of whether the issue falls within academic freedom or tenure rights, the action taken is in violation of the rules governing PPI stated in the Medical Service Plan *(MSP) of the University Medical School and in violation of basic property rights created by state law.*

According to the MSP, while the basis for PPI is subject to negotiation, this negotiation is an annual one. During a given year, the basis for calculation of PPI becomes a property right of the faculty member just as would the amount of "base salary" for any member of the faculty in any school on campus. The Committee is sensitive to the possibility of unforeseen financial problems arising during a fiscal year, but the MSP provides techniques for dealing with shortfalls in revenue and those techniques do not include selective reduction of one person's entitlements or isolation of one person onto a different basis from the rest of the division.

Secondly, placing Dr. Fineman on a strict patient billing formula while allowing the other members of the Division to receive a fixed amount of [annual] PPI income is also against the requirements of the MSP. The MSP provides that every individual's PPI will be based on patient billings of that person, although the unit may "skim" amounts needed to provide additional income for those who generate income without generating patient billings. The Division of Medical Genetics, however, generally is on a system that shares income from a pool of revenue derived from various sources. Dr. Kurveh stated that the entire Division of Medical Genetics is on Plan B, but with the exception of Dr. Fineman, they are paid <u>as if they were on Plan A</u>. According to the MSP, both plans (A and B) are available as options only to <u>entire</u> "departments or sub-units [for example, divisions] within departments;" individual faculty members do not count as "sub-units." It is, therefore, in violation of these regulations to treat Dr. Fineman differently from the rest of his Division. **Moreover, the way that the rest of his Division is in fact paid its PPI accords with neither Plan A nor Plan B [emphasis mine].**

In the Committee's judgment, this treatment of Dr. Fineman was punitive. When asked for an explanation of shifting the method for calculating Dr. Fineman's PPI to a strict patient billing basis, which is not the case for anyone else in the Division, the answer was that Dr. Fineman did not do his share of the work, that he did not participate in the "social contract," and that his behavior was "outrageous." When asked what constituted "outrageous behavior," [Dr. Mamzer] said it was lack of participation. The Committee found these explanations of action vague. The basis for the calculation of his PPI does <u>not</u> reflect his clinical performance, but is a reflection of [Drs. Kurveh's and Mamzer's] judgment on his net worth to the Division. Neither [Dr. Kurveh nor Dr. Mamzer] disagreed with Dr. Fineman's assertion that he would have to see more than 70% of all the Division's patients to earn the same PPI as the other members of the Division are receiving or as he was receiving under the prior arrangement. [How can a faculty member excel in research and teaching if he/she were forced to see more than 70% of his/her division's patients?] *As a reflection of his net pay from Division activities, it constitutes a reduction in his status and requires a "show cause" hearing.*

III. Academic Standing

Finding
The Committee finds that, although some actions were wrong because of their retaliatory nature, the totality of Dr. Fineman's treatment did not constitute a denial of reasonable access to the conditions necessary to discharge his duties as a tenured faculty member. The vote was - Yes: 8; No: 0; Abstain: 0.

Remedy
Dr. Fineman's complaint on this point includes two diverse groups of actions. First are actions taken by [Drs. Kurveh and Mamzer] to redefine his role in the Division of Medical Genetics and increase his clinical activities. Second are actions which, according to Dr. Fineman, were intended to force him to acquiesce in this redefinition or to punish him for not acquiescing. On the first, which Dr. Fineman has claimed amounts to moving him from the research/clinical/teaching [tenure] track to the [non-tenured] clinical track, the AFTC did not accept Dr. Fineman's conjectural theory for the simple reason that nothing has been done to challenge his tenure as a faculty member nor to move him formally to a different position. For this reason, we subsumed this compliant under the broader heading of "a climate suitable for scholarship, research, and effective teaching and learning."

Because we believe these concerns to have been the crux of the disagreement between the parties we wish to make explicit that we do not find in Dr.

Fineman's favor on this matter. Dr. Fineman produced no convincing documentation of his claims that he could not continue his academic career while meeting the same clinical responsibilities as others in his Division. [Drs. Kurveh and Mamzer] argued persuasively he could continue his academic career. We do recognize that Dr. Fineman initially had fewer responsibilities in the Division for clinical care, but changes in the needs of the University can legitimately alter initial working conditions. The changes proposed in this case do not amount to violations of Dr. Fineman's tenure or academic freedom.

The Committee discussion at this point was very involved. Several members of the Committee felt that the treatment of Dr. Fineman was in some deep sense illegitimate and unfair. Although the actions do seem to fall into the category of administrative decisions, and as such are not subject to a "show cause" hearing, nevertheless the manner in which they were taken seemed to the Committee to fall dangerously close to being arbitrary and capricious.

Ultimately, the Committee decided that it was not our business to adjudicate generally the effectiveness of administration within any department or other unit of the university. If a Dean or Department Chair damaged the academic stature of a unit by general personnel actions, then the University would have a problem but that problem would not be within our purview.

Concluding Remarks

Finally, the Committee feels it must make the President aware of [Dr. Mamzer's] uncooperative behavior throughout the course of this hearing, and ask the President to inform [Dr. Mamzer] of the importance of

due process and procedures for protecting the integrity of the University **[emphasis mine]**.

Throughout this appeal, [Dr. Mamzer] has called into question repeatedly the Committee's authority to inquire into these matters. In a letter to the Vice President for Health Sciences dated 18 May 1989, [Dr. Mamzer] wrote:

My principal concern is that once again the Academic Freedom and Tenure Committee has taken unilateral information from an individual who is willing to lie and mislead, and who clearly has personality difficulties. I have checked with the Dean of the medical school; [the Chair of the AFTC from last year] did not discuss this issue with him. [Last year's AFTC Chair] made no attempt to validate Dr. Fineman's contentions with myself or with [Dr. Kurveh]. I am assuming that you also had no discussions with him concerning this issue. The President should censure [last year's Chair] and his committee for their total disdain for fair procedure and due process.

The Committee does not doubt that this represents [Dr. Mamzer's] perception of the facts. It does not, however, accord with the Committee's recollection of the facts. In 1988 - 1989, the Committee was distressed by the recalcitrance which characterized [Dr. Mamzer's] interactions with its representative, [last year's Chair]. [Last year's Chair] reported to the Committee his difficulties in obtaining information or cooperation from [Dr. Mamzer]. This recalcitrance has continued into the hearings on 4 December 1989 and 2 February 1990.

In his response to our request for information (in a letter dated 26 January 1990, two full weeks <u>after</u> the deadline of 12 January 1990 set by the

Committee in order to allow time for Dr. Fineman's rebuttal and a full review of all information by affected parties before the hearing), [Dr. Mamzer] writes:

On the instruction of the Dean of the School of Medicine and the Vice President for Health Sciences, I am responding to your letter of 13 December 1989. I do this over my stated objections to the appropriateness of the inquiry by the Academic Freedom and Tenure Committee. The following response has had no input from my legal counsel because of time constraints of this matter. My appearance before the Academic Freedom and Tenure Committee will depend on the advice of counsel.

Later in the same letter he continues:

It is incredible to me, and to the ultimate detriment of the principles for which a university should stand, that the University faculty governance processes continue to sustain Dr. Fineman's destructive behavior. His definition of an academic career will ultimately be fatal to all of us.

The Committee believes that what will "ultimately be fatal to all of us" is <u>not</u> protecting those rights which are so essential to preserve academic freedom without which no university can survive; further what will be fatal is failure to respect due process necessary for preservation of those rights. **Any respondent has the right to object to the jurisdiction of a tribunal, but [Dr. Mamzer] has shown consistently a total failure to understand the nature and processes of an academic environment.**

The Committee urges the President to accept and act on the findings and remedies contained in this report immediately in order to bring this matter to closure. It has been more than two years since this problem first came to the attention of the AFTC, and we believe it is vital to the health of the University to see it finally brought to a conclusion [emphasis mine].

www.ingramcontent.com/pod-product-compliance
Lightning Source LLC
Chambersburg PA
CBHW032250150426
43195CB00008BA/393